AF836242

Energy Efficient Passive Building: A Case Study of SODHA BERS COMPLEX

Energy Efficient Passive Building: A Case Study of SODHA BERS COMPLEX

G. N. Tiwari
Centre for Energy Studies,
Indian Institute of Technology Delhi, India
gntiwari@ces.iitd.ernet.in

Arjun Deo
Centre for Energy Studies,
Indian Institute of Technology Delhi, India

Vikram Singh
Centre for Energy Studies,
Indian Institute of Technology Delhi, India

Arvind Tiwari
Bag Energy Research Society,
Varanasi, India

now
the essence of knowledge
Boston — Delft

Foundations and Trends® in Renewable Energy

Published, sold and distributed by:
now Publishers Inc.
PO Box 1024
Hanover, MA 02339
United States
Tel. +1-781-985-4510
www.nowpublishers.com
sales@nowpublishers.com

Outside North America:
now Publishers Inc.
PO Box 179
2600 AD Delft
The Netherlands
Tel. +31-6-51115274

The preferred citation for this publication is

G. N. Tiwari, A. Deo, V. Singh and A. Tiwari. *Energy Efficient Passive Building: A Case Study of SODHA BERS COMPLEX*. Foundations and Trends® in Renewable Energy, vol. 1, no. 3, pp. 109–183, 2016.

This Foundations and Trends® issue was typeset in LaTeX using a class file designed by Neal Parikh. Printed on acid-free paper.

ISBN: 978-1-68083-196-2

Editorial Scope

Topics

Foundations and Trends® in Renewable Energy publishes survey and tutorial articles in the following topics:

- Renewable energy technologies
  - Solar thermal
  - Photovoltaic
  - Wind
  - Hydro-electricity
  - Bio-mass
  - Bio-gas
  - Wave
  - Tidal
  - Ocean thermal
- Systems
  - Energy storage
  - Grid integration
  - Conversion technologies
  - Underpinning materials developments
- Non-technical
  - Economic viability
  - Market development
  - Legal and regulatory

Information for Librarians

Foundations and Trends® in Renewable Energy, 2016, Volume 1, 4 issues. ISSN paper version 2328-8892. ISSN online version 2328-8906. Also available as a combined paper and online subscription.

Foundations and Trends® in Renewable Energy
Vol. 1, No. 3 (2016) 109–183

DOI: 10.1561/2700000003

Energy Efficient Passive Building: A Case Study of SODHA BERS COMPLEX

G. N. Tiwari
Centre for Energy Studies,
Indian Institute of Technology Delhi, India
gntiwari@ces.iitd.ernet.in

Arjun Deo
Centre for Energy Studies,
Indian Institute of Technology Delhi, India

Vikram Singh
Centre for Energy Studies,
Indian Institute of Technology Delhi, India

Arvind Tiwari
Bag Energy Research Society,
Varanasi, India

Contents

Abstract

A composite climate case study has been carried out of a newly-constructed, passive-cooled, four-story building in Varanasi (UP), India. This four-story, passively-cooled building is known as SODHA BERS COMPLEX (SBC). In the design of the SBC building, most of the older cooling concepts such as orientation, cross ventilation, day lighting, unglazed Trombe walls, earth sheltering, wind towers etc. were considered. Additionally, a solar water heating system, a roof top PV system and a photovoltaic thermal greenhouse dryer have also been integrated, in order to meet the energy demand of buildings. The total initial embodied energy, CO_2 emission, annual energy saving and CO_2 credit of SBC have all been evaluated. Based on the present study, it has been seen that during harsh summers and cold climatic conditions, the earth sheltering/basement ($\cong$ 28°) and first and second floors ($\cong$ 18–20°) are the most comfortable zones from a passive cooling/heating point of view respectively. It has also been observed that there are approximately 4740 tons of earned CO_2 credit with (i) a payback period of 20 years, and (ii) one unit of energy saving of ₹3 for an average 10° temperature difference between the room and the ambient air temperature.

G. N. Tiwari, A. Deo, V. Singh and A. Tiwari. *Energy Efficient Passive Building: A Case Study of SODHA BERS COMPLEX*. Foundations and Trends® in Renewable Energy, vol. 1, no. 3, pp. 109–183, 2016.
DOI: 10.1561/2700000003.

1

Introduction

The depletion of fossil fuels and climate change needs to be addressed in every sector of life, including building design, which has been continuously increasing since the beginning of industrialization. It is important to mention that these two problems are tightly connected and need to be resolved as a holistic solution. In the United States, there is a 48% energy consumption in the building sector against a 29% energy consumption in the domestic sector in India (CEA 2009). The pattern of energy consumption in buildings is shown in Figure 1.1, which clearly shows that 59% of the energy is consumed in the cooling and heating of the building. Only 12% of energy is required for artificial lighting in India. Pacheco et al. [2012] and Stevanović [2013] have reviewed the work on passive heating/cooling mechanism of buildings which includes the orientation, ventilation, radiative cooling, shape, shade, glazing and phase change materials etc. Recently, Chen et al. [2015] have comprehensively reviewed the passive building design criteria by considering the building layout, envelope, thermo physics, building geometry, air-tightness and infiltration performance, and their effects on energy consumption. This shows the strong requirement to design passive cooled/heated buildings across the Indian continent, from fossil

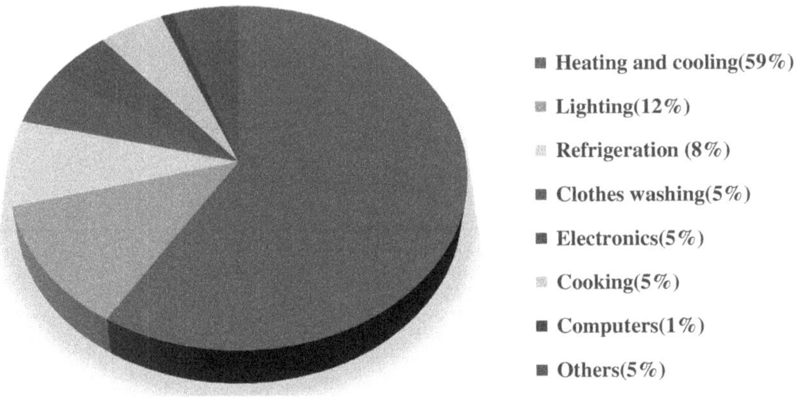

Figure 1.1: Pattern of energy consumption in a residential building (source: Department of energy)

fuel based energy conservation to sustain the climate. Furthermore, the passive design of a building from an energy conservation point of view mainly depends on local climatic condition.

Bansal and Minke [1988] have made an attempt to classify the Indian climatic condition based on a heating and cooling point of view and it is summarized in Table 1.1. However, the proposed classification can be valid globally. For example, the weather in many European countries including the UK, Ireland, Germany, France and Italy fall under the category of cold and cloudy like Srinagar and Jammu. Countries in the Middle East such as Saudi Arabia, Oman, Iran and Iraq, all have a hot and dry climatic condition. Here it is important to mention that the design and performance of buildings from a heating, lighting and cooling point of view will be different for each climatic condition.

From Table 1.1, one can see that the northern part of India, covering a maximum populated area, comes under a composite climatic condition with temperatures ranging between 2 °C and 50 °C. However, a harsh cold climatic condition is only present for a few days during December to January (approximately 30 days) and a warm condition is the norm for about seven months (April to October). Hence, there is a need to design passive buildings in northern India from a cooling

Table 1.1: Criteria for the classification of climatic condition based on monthly average climatic parameters, [Bansal and Minke, 1988].

Climate	Air Temperature in °C	Relative humidity (γ) in %	Precipitation (rain/snowfall) in mm	Number of clear day	Example
Hot and dry (HD)	>30	<55	<5	>20	Jodhpur; Jaipur
Warm and humid (WH)	>30	>55	>5	<20	Chennai; Mumbai
Moderate (MO)	25–30	<75	<5	<20	Bangalore
Cold and cloudy (CC)	<25	>55	>5	>20	Srinagar; Jammu
Cold and sunny (CS)	<25	<55	<5	>20	Leh; Ladah
Composite (CO)	This applies, when six months or more do not fall within any of the above categories				New Delhi; Varanasi

rather than a heating point of view to sustain the climate. As mentioned earlier, there are many passive cooling concepts available in the literature and some of them will be used in the present study by using locally available building materials and skilled manpower.

In the past, energy supply to the building sectors in India is mostly from coal-based power plants. Recently, the government of India is encouraging the use of solar energy in the heating and cooling of buildings by passive and active modes.

2

Background

The **BAG ENERGY RESEARCH SOCIETY** (BERS), registered in 2007 with the government of India, is a non-governmental organization controlled by IIT faculty and former Ph.D. students. It works with public funding for energy education in rural areas of India for sustainable climate. Since there is a lot of scope for energy conservation for a clean environment in passive building design, it was decided by the management committee to incorporate most of the passive concepts used by our ancestors prior to the 1950s in **SODHA BERS COMPLEX (SBC)** at Varanasi (UP) in India. It was inaugurated by Prof. R.K. Shevgaonkar, Director of IIT Delhi in the presence of many international scientists, as shown in Figure 2.1, on February 8, 2012. Prof. G.N. Tiwari discussed passive building in the presence of Prof. Brian Norton, Guest of Honor, during the inauguration. The design details of SBC will be discussed in the next section.

Figure 2.1: Photograph of SODHA BERS COMPLEX (SBC) during inauguration by Director, IIT Delhi.

3

Design of SBC

SBC has been built on a land area of 780 m^2. This is a four-story building. The typical plan for the basement, ground floor, and first and second floors is briefly discussed in the following sections:

3.1 Basement

The basement has been made using an earth shelter concept in order to have a sufficient cooling effect. The open area was used for wind tower 1 (WT-I), located in a southerly direction, which provides day lighting in the basement. Glazed ventilators were also provided at the top of the east and north side walls of the basement for cross ventilation. WT-I allows hot air to pass through it during the summer period. The east and west walls have a provision for cabinets (almirah) for storage purposes as well as to reduce heat loss (U value) from room to side. The basement is mostly used for community services. The isometric view of basement has been shown in Figure 3.1.

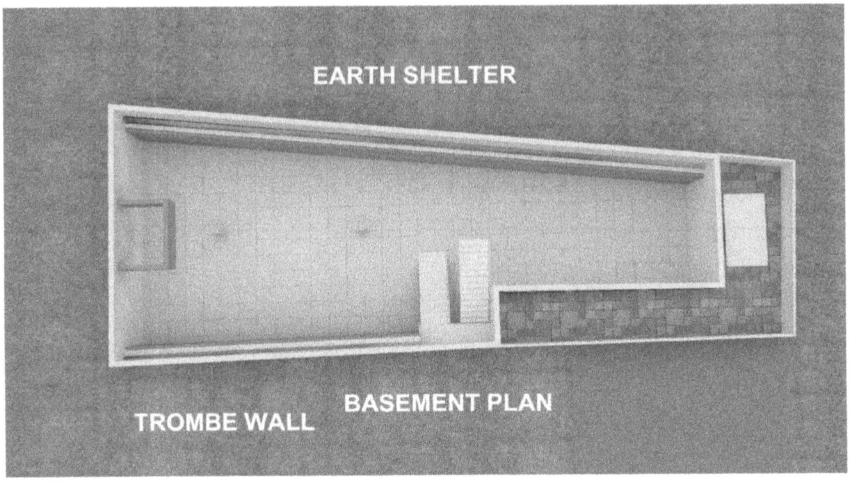

Figure 3.1: Isometric view of basement showing position of WT-I and cavity wall.

3.2 Ground floor

The ground floor plan has the provision of two wind towers. This is in addition to sufficient cross-ventilation from a north to south direction due to a unique natural wind channel from the east, as the building is located at a dead end of the road, which cools the rest of whole building. This ground floor is used for religious functions. There is no partition wall due to the nature of its application, therefore providing a natural passage for first and second floor ventilation.

3.3 First and second floor

The rooms of the first and second floors of the building have a clear height of 3.6 m from a cooling point of view. The top portion of the corridor partition walls is open for proper day lighting and the circulation of cool air from one room to another. In this case, the hot air near the roof is carried away due to cross ventilation. This happens because of the movement of hot air in an upward direction from the lower portion. It is also important to mention that the south facing wall has a minimum exposed area in comparison with the east and

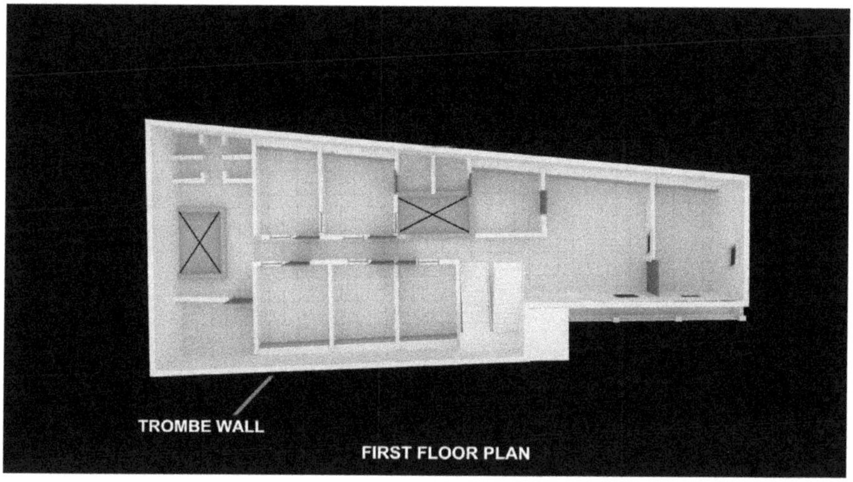

Figure 3.2: Top view of the first and second floor showing the position of WT-I and WT-II and cavity wall.

west walls which minimize the role of heat fraction from the exposed south interior wall. The typical layout plan for first and second floors is shown in Figure 3.2, which has the following configurations:

a. Cross-ventilation through the corridor;

b. Two wind towers;

c. Open ventilation of parallel wall in corridor for the exchange of air as well as cooling;

d. Exposed wall has air cavity through the wooden cabinet (almirah) to reduce the U-value.

These floors are used for residential purpose.

The construction details are as follows.

4

Construction of SODHA BERS COMPLEX

4.1 Foundation

According to the site's ground profile, the basement was built to avoid the land filling cost, as well as to get the bearing capacity of the soil as per design requirement. A minimum foundation depth for non-load bearing wall was given as 0.91 meter stepped footing while the load bearing column foundation was 1.50 meters trapezoidal footing below the existing ground level (G.L.). Other details are shown in Figure 4.1.

4.2 Column

We have provided three types of columns according to the calculated load pattern of the building as shown in Figure 4.2. We curtailed the reinforcement as well as the size of each floor accordingly, decreasing the load to minimize the cost and embody energy simultaneously. The quantity of re-enforced concrete and cement (RCC) used was approximately 42 m^3.

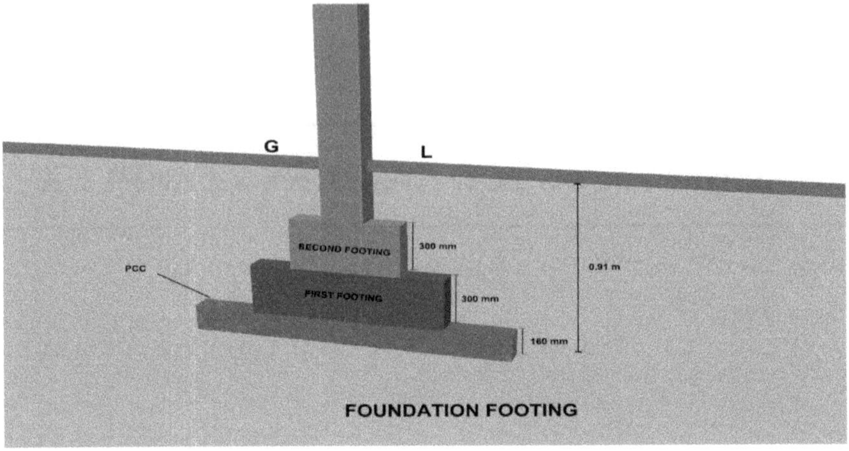

Figure 4.1: The dimension in mm of the foundation for the basement (GL and PCC stands for ground level in the basement and pozzolona portland cement respectively).

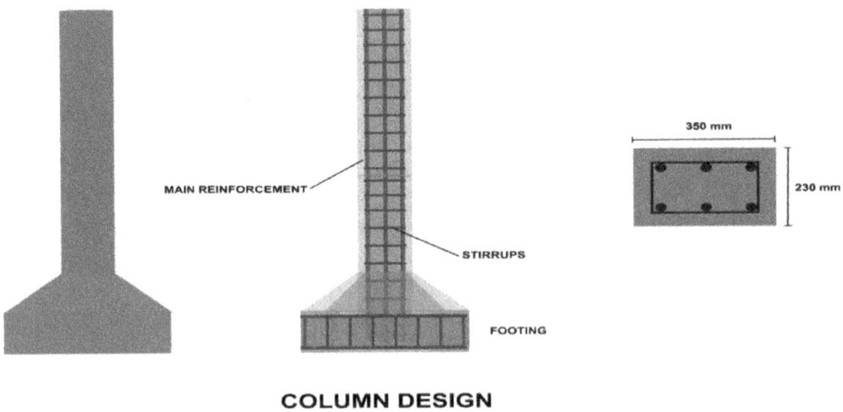

Figure 4.2: View of the column design.

4.3 Beam and Slab

According to the span and load we designed three types of beams. The main beam has the detail reinforcement and size as shown in Figure 4.3. The design is made as per the T-beam theory and the quantity of reenforced concrete and cement (RCC) was nearly 107 m^3.

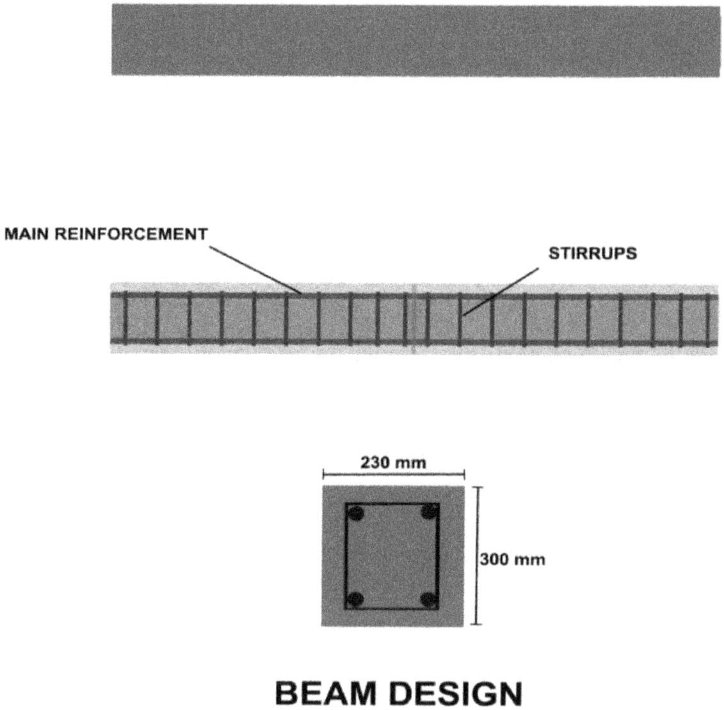

Figure 4.3: The design of the beam used for SBC.

4.4 Outer Walls

The brick outer walls have the thickness of 0.115 m, with a depth of almost 0.550 m for the wooden cabinet. The inner wooden cabinet creates an air cavity to minimize the U-value of the vertically exposed wall to reduce heat gain from the outside environment. All of the inner walls have the same thickness with an opening of 0.75 m at the top for ventilation as well as day lighting. At night corridor illumination provides plenty of light within the rooms. We have saved about 340 tons of brick over a conventional construction.

4.5 Roof top PV system

There are 12 PV modules (900 Wp), each of 75 Wp, as shown in Figure 4.4, to provide emergency light to complex in the absence of grid power. There are four semi-transparent and eight opaque PV modules. All 12 PV modules are connected in series to provide a 12 V and 46 A current. Two 12 volt batteries have been used to store electrical power for use during the night. There has been an annual electrical energy saving of about 1124 kWh since 2012, Deo et al. [2015]. The embodied energy of the roof top power system has been evaluated at 8493.16 kWh. The energy matrices, Deo [2013], for the roof top are given by:

$$\text{Energy Payback Time (EPBT)} = \frac{\text{Embodied energy}}{\text{Annaul electrical energy}} = \frac{8493.16}{1124.00} = 7.56 \text{ years} \tag{4.1}$$

$$\text{Electricity Production factor (EPF)} = \frac{\text{Electrical energy generated in whole life}}{\text{Embodied energy}} = \frac{1124 \times 30}{8493.16} = 3.97 \tag{4.2}$$

Figure 4.4: Roof top PV system of 900 W_p.

and

$$\begin{aligned}&\text{Life cycle conversion efficiency (LCCE)}\\&=\frac{\text{Electrical energy generated in whole life–Embodied energy}}{\text{Yearly average solar radiation} \times \text{life of system}}\\&=\frac{1124 \times 30 - 8493.16}{624.37 \times 7.32 \times 30} = 0.18 \qquad (4.3)\end{aligned}$$

The average yearly sunshine hour and typical lifetime of the roof top PV system has been considered as 7.32 hours and 30 years respectively.

From the above energy matrices, one can see that the roof top PV system can be feasible without the degradation of the PV module and land cost of the building which can be justified.

4.6 Solar water heating system

There are two solar water heating systems, each having a capacity of 200 liters to meet the requirement of hot water to the building. One is based on a conventional flat plate collector (FPC) and the other is an evacuated tubular collector (ETC). The kitchen and bathroom on each floor have been provided with one terminal of hot water. The hot water is mainly used for the cooking and washing of the resident's clothes. It has been noted that an FPC water heating system is 40% more expensive than an ETC one. However, the maintenance of an ETC water heating system is expensive in the case of an evacuated tube breaking. The performance of an ETC water heater is much better (about 20%) than an FPC. There is an annual saving of 2000 kWh thermal energy due to hot water being available to the building. In this case, the energy payback time is always between 2 and 4 years, Tiwari [2006].

In following section, we will discuss the review of some of the cooling concepts that were used in SBC for a composite climatic condition as mentioned in Table 1.1.

5

Passive Cooling Concepts

5.1 Trombe wall

A solid thick concrete/brick wall, as shown in Figure 5.1a, reduces (i) the heat flux coming from exposed outer wall to inside room, and (ii) the fluctuation of room air temperature due to a small overall heat transfer coefficient (Eq. 5.1a) and the high heat capacity of the exposed wall. For an exposed heavy structure wall for a passive building, one can have (a) a bare exposed surface (unglazed Trombe wall, Figure 5.1b) (absorptivity, $\alpha \leq 0.4$) for the thermal cooling of the building, and (b) a blackened (absorptivity, $\alpha \geq 0.9$) exposed surface (glazed Trombe wall, Wikipedia) for the thermal heating.

This concept was proposed and patented by Trombe in 1972 and the results based on this concept were published in 1974 (Trombe, 1972 and 1974). The unglazed heavy structure wall consists of stone, bricks, reinforced concrete cement (RCC) and mud. It is regarded as one of the green building envelopes (passive concept), which is sustainable for cooling (unglazed) and heating (glazed) to address the environmental and energy crises worldwide. In the past, an unglazed heavy structure wall/roof was used frequently all over world, particularly in India. Figure 5.1a shows a part of one fort in India with the ventilation

Figure 5.1a: Photograph of a fort in North-India constructed a long time ago by using thick brick wall for cooling during summer.

arrangement shown above the door. The temperature inside has been observed as comfortable throughout the year. The fort is over 500 years old. Furthermore, all exposed and partition solid walls of the buildings in Banaras Hindu University (BHU), Varanasi, have a thickness of over 0.60 m and these buildings provide better thermal comfort in summer conditions, with less fluctuation in room air temperature (~25–28°C) against the outside temperatures of (~45–48°C). The buildings in BHU employ such techniques that are almost 100 years old and these are being used for teaching and research.

Gan [1998] has carried out parametric studies of Trombe walls for the passive cooling of buildings, while Kruger et al. [2013] has evaluated the performance for subtropical locations. Stazi et al. [2012] have discussed the management of Trombe walls in summer conditions, and the optimization of energy and environmental performance of passive Trombe walls has been carried out by Bojic et al. [2014]. Saadatian et al. [2012] have reviewed the application of Trombe walls in buildings,

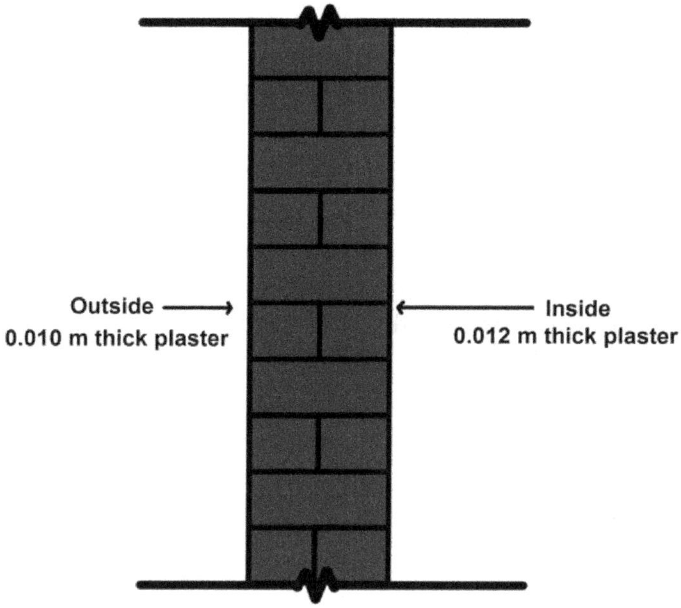

Figure 5.1b: Unglazed heavy structure Trombe wall used for summer climatic condition.

including Trombe wall configurations and technology with its advantages and disadvantages. Ozel [2013] has studied the placement of optimum insulation thickness over Trombe walls on a cooling transmission load for a hot climate. Tummu et al. [2013] have studied the different levels of insulation and its placement on exterior and interior surfaces, both on walls with and without windows. They have concluded that for a Thailand climatic condition (warm and humid condition, Table 1.1), insulation placed on the interior wall surface improves the wall thermal performance and is much better than the exterior placement for air-conditioned buildings. Wang et al. [2007] have noted that the insulated exterior walls of residential buildings in a cold climate are cost-effective. Al-Sanea and Zedan [2011] have shown that the placement of insulation is better inside and outside (exposed) for heating and cooling respectively, for a given total thermal resistance. This is in agreement with results reported earlier by Sodha et al. [1979a,b].

For the thermal heating of buildings, some other modified Trombe walls, namely the PV-Trombe wall, PCM Trombe wall and air vent Trombe wall have also been studied by Jie et al. [2007], Tyagi and Buddhi [2007] and Liu et al. (2014).

Up to 1950 however, the concepts of thick solid walls (Trombe wall) (>0.60 m) were very common for cooling the buildings in India during the summer months. The solid mud wall used would have a minimum U-value but require annual maintenance. At present, most of the buildings in India are using brick walls of about 0.115 m (a higher U-value) including partition walls for a minimum annual maintenance on the cost of using expensive grid power for good thermal comfort in summer as well as winter. This phenomenon is common due to the cost effectiveness of brick materials in housing construction.

On the basis of the review carried out above, one can even improve upon the existing walls for cooling by the following techniques:

(i) By using appropriate locally available waterproof insulating materials to the exposed (exterior) wall to reduce the U-value, Sodha et al. [1979a,b], Wang et al. [2007] and Al-Sanea and Zedan [2011].

(ii) By using locally available insulated (wooden/plastic sheet) cabinet (almirah) on the interior of the exposed walls of the building. The use of such interior insulated cabinets for clothes etc. ensures a much lower U-value. Hence this concept has been used in SBC.

The total thermal transmittance, also known as U-value, for single and multilayered Trombe walls can be expressed as follows:

(a) For single-layered unglazed Trombe wall, Tiwari [2002]

$$U = \left[\frac{1}{h_o} + \frac{L}{K} + \frac{1}{h_i}\right]^{-1} \tag{5.1a}$$

where $h_o = 5.7 + 3.8\,\text{V}$ is convective and radiative heat transfer coefficient (W/m^2°C) from exterior wall surface to ambient air facing wind speed (V) and $h_i = 2.8$ W/m^{2}°C is convective heat transfer coefficient from interior wall surface to room air.

For brick wall thickness (L) = 0.60 m and 0.115 m with thermal conductivity of K = 0.69 W/m °C, and zero wind speed, the U value from Eq. (5.1a) can be obtained as follows:

$$U = [1.40]^{-1} = 0.71\,\mathrm{W/m^2\,°C} \quad \text{for L} = 0.60\,\mathrm{m}$$
$$= 1.43\,\mathrm{W/m^{2}°C} \quad \text{for L} = 0.115\,\mathrm{m}$$

It can be seen that there is increase in U value by 101% by reducing the thickness of a solid wall from 0.60 to 0.115 m. Hence one can conclude that in order to thermally cool a building, a solid wall thickness plays an important role in reducing the incoming flux into room.

(a) For multilayered-layered Trombe wall

$$U = \left[\frac{1}{h_o} + \sum_{j=1}^{N} \frac{L_j}{K_j} + \frac{1}{h_i}\right]^{-1} \tag{5.1b}$$

If a solid brick wall of thickness L = 0.115 m is covered with waterproof insulated material of thermal conductivity K = 0.04 W/m °C and thickness L = 0.075 m from either exterior or interior wall, then Eq. (5.1b) can be used to evaluate U value for waterproof insulated wall as follows:

$$U = [0.175 + 0.167 + 1.875 + 0.357]^{-1} = [2.7405]^{-1} = 0.38\,\mathrm{W/m^{2}°C}$$

One can further note from above calculation that the U value of a waterproofed insulated wall of brick thickness of 0.115 m is half value in comparison with a brick wall of thickness 0.60 m. This has been supported by many researchers, Wang et al. [2007] and Al-Sanea and Zedan [2011].

Now, the rate of heat transfer from an exposed wall to room air can be expressed as follows:

$$\dot{Q}_u = UA_w(T_{sa} - T_r) \tag{5.2a}$$

where $\dot{Q}_u$ is the heat transfer in W, U is the thermal transmittance in W/m^{2}°C, T_r is the room air temperature, °C, on one side of the structure, A_w is the exposed (exterior) wall area in m^2 and T_{sa} is the solair temperature, °C, on the exposed (solar radiation) side of wall.

An expression for solair temperature (T_{sa}) is given by

$$T_{sa} = \frac{\alpha}{h_o} I(t) + T_a - \frac{\varepsilon \Delta R}{h_o} \tag{5.2b}$$

where α and ε are the absorptivity and emissivity of the exposed surface of the wall, $\Delta R = 60$ W/m^2 is the longwave length radiation exchange between the ambient and sky and it is zero for vertical wall. I(t) is the incident solar radiation on the wall in W/m^2.

Furthermore, solair temperature (T_{sa}) mainly depends on solar intensity. I(t) and ambient air temperature T_a, which are periodic in nature and hence solair temperature, can also be considered as periodic. Due to this, the behavior of room air temperature is considered as a periodic function. Mathematically, solar air temperature can be expressed as follows:

$$T_{sa}(t) = T_{sa0} + Real \sum_{n=1}^{\infty} T_{san} \exp(-i\phi_n) exp(in\omega t) \tag{5.2c}$$

where $T_{sa,0}$, $T_{sa,n}$ and ϕ_n are Fourier coefficients and can be evaluated for known values of solar intensity and ambient air temperature. The 'n' is number of harmonics.

During off sunshine hours (night), Eq. (5.2a) is reduced to

$$\dot{Q}_u = UA(T_r - T_a) \tag{5.3a}$$

The tradition in India until recently (~60 years back) was to use the concept of unglazed Trombe walls (thick/heavy structure walls) by using locally available materials to construct the residential houses for different climatic conditions as mentioned earlier. Figure 5.2 shows a housing complex built for a composite climate in Jaipur, Rajashtan, India, see Table 1.1 for thermal cooling. Locally available mud has been used for the thick walls (lower U-value) in all houses by using the concept of unglazed Trombe walls with a white color for minimum absorption of solar radiation for further cooling. The roof is made up of bamboo and dried grass which is also cost effective, along with thermal comfort throughout the year for a hot and dry climate. Such houses were most appropriate for rural housing in India.

Figure 5.3 shows another rural house built for hot and humid conditions along the coastal belt in Chennai, India. One can see that the house has the facility of day lighting through the multilayered roof (small U value) as well as a ventilation system in the wall. However,

Figure 5.2: Photograph of a housing complex with heavy structure walls (lower U-value, Eq. 4.1a) for cooling in summer in Jaipur.

it is important to mention that the outer exposed wall is not white as shown in Figure 5.2. It has been observed that the room air temperature in Figures 5.2 and **??** is $7-8°C$ below the outside ambient air temperature. This is due to the storage effect of unglazed Trombe (thick) walls. In both cases, locally available materials have been used to make the house cost effective. Both houses were maintained by unskilled labour.

The partition walls are generally used internally and these are not exposed to solar radiation. The walls dividing the two adjacent rooms inside the building have similar room air temperature. Even for a larger U value, there is not much heat transfer exchange between the two rooms. This is more realistic if there is a direct connectivity through the top opening at the partition wall. In this case, the thickness of the partition wall can be considered as 0.115 m. The U value for the

Figure 5.3: Photograph of a rural house with a heavy structure wall for hot and humid conditions.

partition wall can be evaluated from Eq. 5.1a as:

$$U = \left[\frac{1}{2.8} + \frac{0.115}{0.69} + \frac{1}{2.8}\right]^{-1} = [0.3571 + 0.1667 + 0.3571]^{-1}$$
$$= [0.88]^{-1} = 1.14\,\mathrm{W/m^{2\circ}C}$$

Here the values of h_o and h_i are the same due to the non-exposure partition walls to the wind.

The rate of thermal energy exchange ($\dot{Q}_{ex}$) in W/m^2 between the two adjacent rooms through the partition wall can be written as follows:

$$\dot{Q}_{ex} = UA_p(T_{r1} - T_{r2}) \tag{5.3b}$$

Here T_{r1} and T_{r2} are the room air temperature of the two adjacent rooms, and A_p is the area of the partition wall (m^2). For any known values of (UA_p), the numerical value of $\dot{Q}_{ex}$ will be insignificant for least a (lower) value of $\Delta T_r = (T_{r1} - T_{r2})$ in comparison with an exposed (exterior) wall.

Since the thickness of the partition wall has been considered as 0.115 m with an opening at the top in SBC, this effect has been neglected in the thermal modeling and experimental validation.

5.2 Air cavity cabinet wall

In order to reduce the cost of treated insulated walls, a concept of an air cavity integrated Trombe wall has been introduced by many scientists. Mahlia and Iqbal [2010] have observed that energy consumption and emissions can be reduced by 65–77% by introducing an optimal thickness of different insulation materials with air gaps of 2–6 cm. An experimental study has been carried out by Charde and Gupta [2013] for a proposed static sunshade with a brick cavity wall for a composite climatic condition. They found that the proposed static sunshade brick projections and brick air cavity wall has been effective in keeping indoor air temperature less in summer mornings and more during winter nights. It is useful for energy conservation in buildings as per seasonal needs. The effects of circular air cavity on the heat storage capacity of the concrete walls have been studied by Zhang and Wachenfeldt [2009]. Antar [2010] has seen the effect of radiation in a rectangular shape brick wall on the performance of room air temperature. Similar studies have also been carried out by Sodha et al. [1979a,b] and Singh et al. [1981].

If one can make a cabinet of 0.550 m along an interior exposed unglazed Trombe wall for keeping usable items e.g. clothes, see Figure 5.4a, the expression for U becomes as such:

$$U = \left[\frac{1}{h_o} + \sum_{j=1}^{N} \frac{L_j}{K_j} + \sum_{j=1}^{N} \frac{1}{C_j} + \frac{1}{h_i} \right]^{-1} \tag{5.4}$$

where $C_j = 4.5$ W/m$^{2\circ}$C is the thermal air conductance of air cavity of thickness 0.550 m.

The U value of a solid brick wall (L = 0.115 m) with a wood cabinet (K = 0.14 W/m °C) with a depth of 0.550 m and thickness of 0.18 m

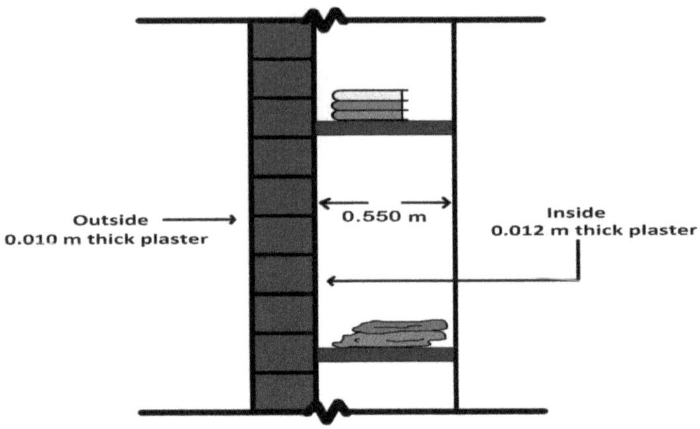

Figure 5.4a: Exposed unglazed Trombe wall with cabinets for storage.

Figure 5.4b: Photograph of inner cabinets with a thin exposed unglazed Trombe wall in SBC.

is obtained from Eq. 5.4 as:

$$U = [0.1754 + 0.167 + 0.222 + 1.286 + 0.3571]^{-1}$$
$$= [2.2075]^{-1} = 0.45\,\mathrm{W/m^{2\circ}C}$$

For SBC, we have considered the configuration shown in Figure 5.4b for the above calculated U value.

If the brick wall is further waterproof insulated, then the U value is reduced to 0.27 W/m$^{2\circ}$C.

Figure 5.4b shows the photograph of a constructed exposed brick wall (0.115 m) with a single (cabinet) air cavity (0.550 m) along with a storage facility at SBC. The cost of the exposed brick wall (0.115 m) with a single (wooden cabinet) air cavity (0.550 m) is about ₹2770/- per m^2 ($44.23), which is about same as the cost of a 0.60 m thick cemented brick wall (₹2767/-per m^2).

5.3 Ventilation/Infiltration

Cao et al. (2014) have reviewed the work on the performance of different ventilation configuration and airflow distribution systems in the building. Recently, research work on wind-driven cross-ventilation has also been carried out by many scientists, Chu and Bo-Fan [2014], Faggianelli et al. [2014] and Schulze and Eicker (2013). In India, most urban houses have a provision of clerestory windows (Rosan dan) near the roof as shown in Figure 4.2. This is to remove the hot air near the roof in natural mode. Hot air has a tendency to move in an upward direction due to its low density and therefore accumulates near the roof. If cooler air is driven inside the room either through the doors or windows, then it becomes easier to remove the hot air through the clerestory window, see Figure 5.5a. The clerestory window should be open in summer for passive cooling and closed during winter for passive heating purposes. In SBC there is a provision for clerestory windows as well as natural cross-ventilation in the basement, first and second floors as shown in Figure 4.2. In this case, there is a movement of natural wind from the north to the south through the opening provided on the east side. There is a bigger heat transfer from the inner wall surface to flowing air through the room due to a large inner wall surface area and therefore a cooling effect is observed inside the room. The heat transfer coefficient (h_i) depends on a variable air velocity (v) through the corridor which can be considered as follows:

$$h_i = 2.8 + 3v \tag{5.5}$$

Figure 5.5a: View of a clerestory window (Rosan dan) in New Delhi for day lighting and ventilation for passive cooling.

Here the radiation term has not been considered due to a small temperature difference between the two walls. The numerical value of cross-ventilation (v) is measured in the corridor and the average value is considered.

Furthermore, the heat transfer rate $\dot{Q}_{v/i}$ (W) can be evaluated through a clerestory window in the room by using the following expression:

$$\dot{Q}_{v/i} = 0.33NV_0(T_r - T_a) \tag{5.6}$$

where N is the number of air change from room air to outside, and V_0 is the volume of the corridor/room through which hot air is withdrawn/removed. The same concept along with unglazed Trombe (thick brick) wall has been shown in Figure 5.5a in old building in New Delhi for passive cooling and heating. Natural ventilation depends on the difference between the room air and outside air temperatures. Figure 5.5b shows the provision of ventilation through a corridor of SBC.

Figure 5.5b: View of natural cross ventilation on the first and second floor at SBC.

5.4 Wind towers

The concept of wind towers in a building has been traditionally used for passive cooling for a harsh warm climate. These are massive structures connected to each floor of the building that capture the local wind mainstream to force fresh air to circulate inside and provide thermal comfort to occupants. Recently, a preliminary design strategy for a steady stack ventilation in multi-story atrium buildings by using a simplified mathematical model has been developed by Acred and Hunt [2014] for unidirectional air flow. A similar study has been carried out by Reyes et al. [2013] and Haw et al. [2012]. Based on a review in the area of the wind tower and the experience in the area of passive building for thermal cooling, we made a recommendation for two wind towers at SBC. The first wind tower (WT-I), as shown in Figure 5.6a, is located at the southern end from the basement to the top floor and is to serve two objectives, namely (i) to provide day lighting to the basement, and (ii) create a wind draft to move the hot air from the basement to the

Figure 5.6a: View of wind tower 1 (WT-I) from the basment at SBC.

outside for cooling under natural as well as forced mode. There are provisions of four small fans, each having a 20W capacity at the upper portion of dryers constructed at the top of WT-I. This wind tower is connected to a wind corridor on the first and second floors. The second wind tower (WT-II) is in the middle of the building and connected to only the first and second floors. WT-II, see Figure 5.6b, also provides the day lighting to each connected floor and removes the hot air from each floor to the top of the building by either forced or natural mode. On the top of WT-I, there is a semi-transparent photovoltaic thermal (PVT) greenhouse dryer with a capacity of 100 kg as shown in Figure 5.7. The semi-transparent PV module was considered because of its higher electrical efficiency, Tiwari et al. [2011], and being able to get the light intensity through the non-packing area of PV module for day lighting through multi-reflection from the walls of WT-I. The PVT dryer is used to dry all locally available medicinal/vegetables crops under forced mode, which are then used by the residents in the complex. During the summer months, the fan is operated for both the

Figure 5.6b: View of wind tower 2 (WT-II) from the ground floor at SBC.

fast drying of agro products and to suck hot air from the bottom to the outside. There are eight semi-transparent PV modules (280 Wp), each 35 W_p, that have been used to provide electrical power to four fans with a rated capacity of 20 W at the north side of the dryer. The PV modules are connected to a battery through a charge controller to store electrical power to provide the light on the staircase during the night. There is an annual electrical energy saving of 180kWh, Nayak and Tiwari (2011), Nayak et al. [2011]. The top of WT-II is also covered with a transparent plastic sheet, as shown in Figure 5.6b, to protect it from the rain and heavy dust during the summer. For cooling in summer and heating in winter, the side windows are opened and closed respectively.

5.5 Earth air heat exchanger

The concept of an earth air heat exchanger is well known for its ability to heat and cool buildings in the past. In this case, the spiral heat

Figure 5.7: Side view of the PVT greenhouse dryer on top of WT-I with four fan on top of the dryer.

exchanger is (preferably) made up of a rigid plastic tube. The spiral heat exchanger is connected in parallel, if needed, to increase the volume of hot/cool air. The heat exchanger is buried ≥3–4 m deep as the ground beyond this depth provides a constant temperature (≅ 25°C). This is valid for all soil conditions, Tiwari [2002]. In this case, hot air from outside the room can flow through the buried heat exchanger and is cooled due to the heat transfer from hot air to cool surface through the heat exchanger pipe and vice versa for cool air to be heated. The cool air is then passed to the room of the building to be cooled. In past years this concept has been used in some of the old buildings. There is a vast amount of literature available from which an earth air heat exchanger (EAHE), integrated to a building for heating as well as cooling, can be designed, [Tiwari, 2002; Dong and Jinpeng (2014); Li et al., 2014, Maerefat and Haghighi, 2010, Peretti et al., 2013 and Bisoniya et al., 2013]. Recently, Peretti et al. [2013] have reviewed the work on a heat exchanger which includes work carried out by the research groups of Prof. M.S. Sodha and Prof. G.N. Tiwari.

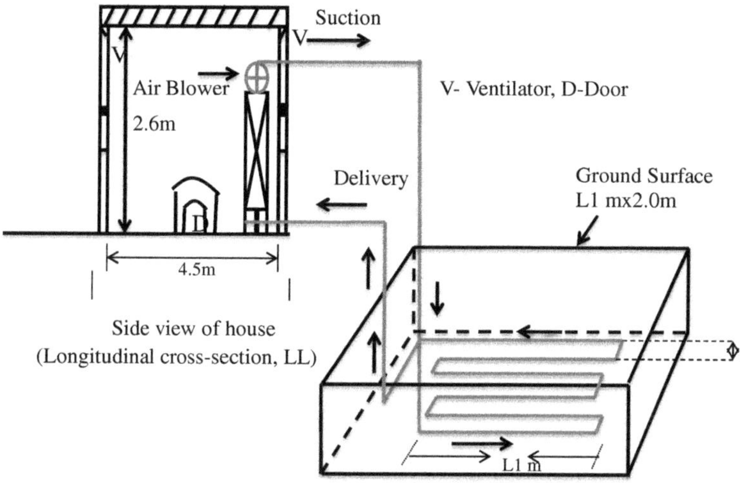

Figure 5.8: An earth air heat exchanger integrated to a room.

Figure 5.8 shows a schematic diagram of a spiral type EAHE integrated into a room for thermal heating/cooling. The heat exchanger is buried at a constant ground temperature i.e. T_0. It may be equal to an average annual ambient air temperature at a depth of 4 m under the ground as mentioned earlier.

Following Tiwari [2002] and Tiwari et al. [2014], the outlet air temperature (T_{fo}) at the end of the heat exchanger length (L) buried inside the ground at temperature T_0 is given by:

$$T_{fo} = T_0\left(1 - e^{-\frac{2\pi r h_c L}{\dot{m}_a C_a}}\right) + T_{fi}e^{-\frac{2\pi r h_c L}{\dot{m}_a C_a}} \tag{5.7}$$

where $r = \frac{d}{2}$ is the radius of heat exchanger, $\dot{m}_a$ and C_a are mass the flow rate and specific heat of air, and h_c is the convective heat transfer coefficient from the inner surface of the heat exchanger to the flowing air.

The above equation has been solved by using boundary conditions at x = 0, $T(x) = T_{fi}$, the initial air temperature which may be hot in summer and cool in winter season. The variation of the outlet air temperature with the length of the heat exchanger is shown in Figure 5.9. It can be seen that for thermal cooling there is a decrease in

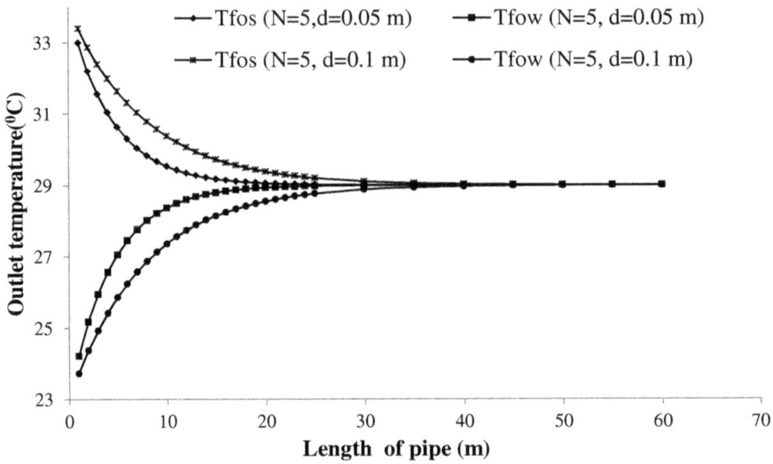

Figure 5.9: Outlet air temperature with length of heat exchanger for different number of air change (N) and diameter of heat exchanger (d).

temperature with a length of up to 25 m of the heat exchanger and vice versa for thermal heating. In this case, the length of the heat exchanger is optimized. The suffixes 's' and 'w' stands for summer and winter respectively. If one needs additional hot/cool air then such a heat exchanger is connected in parallel.

The rate of thermal energy gain by the air from the EAHE is given by:

$$\dot{Q}_{\mathrm{u}} = \dot{m}_{\mathrm{a}} C_{\mathrm{a}} (T_{\mathrm{fo}} - T_{\mathrm{fi}})$$

or

$$\dot{Q}_{\mathrm{u}} = \dot{m}_{\mathrm{a}} C_{\mathrm{a}} (T_{0} - T_{\mathrm{fi}}) \left[1 - e^{-\frac{2\pi r h_{c} L}{\dot{m}_{a} C_{a}}}\right] \tag{5.8}$$

Figure 5.10 shows the variation of the rate of thermal energy with the length of the heat exchanger, Tiwari et al. [2014]. It is to be noted that for thermal cooling, the rate of thermal energy is negative and vice versa for thermal heating. By using an EAHE, fossil fuel can be saved to protect the environment. The number of heat exchangers depend upon the size of the room and number of air changes (N) required to achieve a comfortable temperature range.

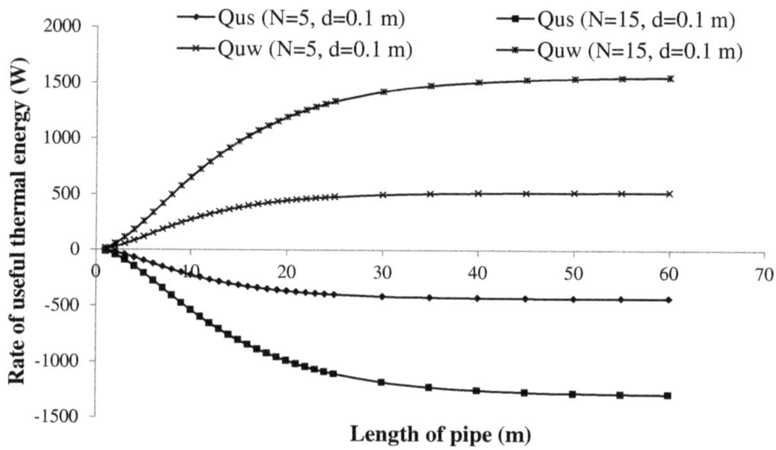

Figure 5.10: Available rate of thermal energy with length of heat exchanger for different number of air change (N) and diameter of heat exchanger (d).

At SBC we have not used an EAHE; however, a provision has been made to integrate the same in the future.

5.6 Height of a room

In the past, it was general practice to keep the height of the room as 6–7 m in a single story building with a provision of open/glazed windows near the roof as a clerestory window, as shown in Figure 5.5a. In this case, the open clerestory window provides an escape route for hot air from inside the room either externally or internally. This also provides day lighting into the room. It is important to mention here that all government buildings across India constructed during British rule prior to 1947 have followed the concept of keeping the building height to more than 6–7 m with clerestory windows for thermal cooling and day lighting. Examples of such buildings are (i) Banara Hindu University (BHU) in Varanasi (UP) which was built from 1960 to 29130; (ii) the North and South blocks (cabinet secretariate), Parliament, New Delhi which were built from 1910 to 1927; and (iii) the Taj Mahal which was built in the year 1652. In all cases, the concept of bare thick Trombe walls has been used for thermal cooling as mentioned earlier.

Furthermore, it is to be noted that all such buildings are either of single or double story due to less population density.

Currently, the construction of single or double-story buildings has changed that of a high rise variety due to a fast growing population in both developing and under-developed countries. Therefore the height of the ceiling in a high rise building needs to be changed to a lower value for an economical viability from a user point of view. The current trend for the ceiling height in high rise buildings is about 9–10 feet (3m) without clerestory windows, which is half of the recommended height of 6–7 m. With this 3 m height and together with a glass window, the rooms are uncomfortable for a hot climate, see Table 1.1. This may be better for a colder climate due to a lower heat loss from the inside to outside ambient air. In a high rise building, a glass window instead of a clerestory window is provided, which isn't appropriate from a security point of view.

In the present building (SBC), the height of the ceiling in each room except the basement has been kept to 3.6 m with the provision of a clerestory window. Furthermore, each room is interconnected through an opening at the top of the partition wall so that hot air can be mixed and then carried away through the wind tower via cross-ventilation. The additional cost due to the increased height in comparison with the present trend could be about 5–10% higher but it saves a lot of conventional energy due to the cooling effect.

5.7 Green roof

There are many techniques to make a green roof. In this case, an exposed surface of multilayered roofing should be treated as follows:

(a) Vegetation, insulation (false ceiling) and an air cavity for ventilation.

For thermal cooling, the placement of insulation either below or above is important. However, the insulation for a roof should be rigid and waterproof if placed outside, as shown in Figure 5.11, unlike the wall configuration. For a further reduction of heat flux, one can use vegetation

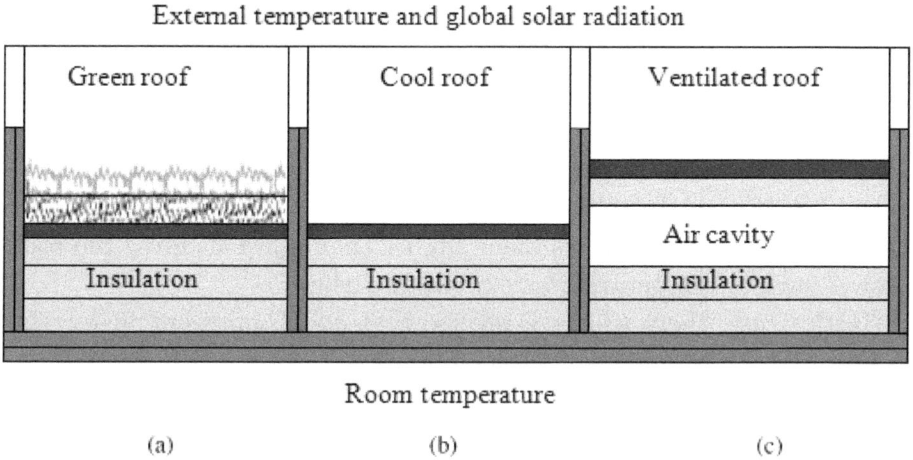

Figure 5.11: The cross-sectional view of configuration of (a) green roof, (b) cool roof, and (c) ventilated roof.

over the insulated roof as suggested by Spala et al. [2008] and Liu and Baskaran [2003] etc.

(b) Wetted roof

In the case of a wetted roof, the availability of water in an arid region is an important issue due to cooling the roof during the summer season. If water is available then flowing water or a pond system over the roof can be adopted. A wetted roof surface provides the evaporation (mass transfer) from the roof surface to air due to unsaturated ambient air. This is more effective in arid (dry) regions due to the higher difference in dry and wet bulb temperatures. It prevents the heat transfer from the exposed surface to the inside room and simultaneously provides the evaporative cooling, Sodha et al. [1980a,b]. Therefore, the wetted roof temperatures are lower than the ambient air temperature.

In the above mentioned cases, there is a strong need for water to be used either with vegetation or a wetted roof during the summer months. This can be achieved by having a rainwater harvesting arrangement in the building.

In this case, an expression for solair temperature becomes:

$$T_{sa} = \frac{\alpha}{h_1} I(t) + T_a - \frac{\varepsilon \Delta R}{h_1} \tag{5.9}$$

where h_1 is the total heat transfer coefficient from a wetted surface to ambient air which includes the evaporative heat transfer coefficient. The numerical value of h_1 is much higher than h_0. This gives the roof a lower surface temperature as explained earlier.

In this case, there will be a reverse heat flow from room air to roof surface and it will be expressed as follows:

$$\dot{Q}_{ex} = U A_p (T_r - T_{x=0}) \tag{5.10}$$

where U can be expressed as

$$U = \left[\sum_{j=1}^{N} \frac{L_j}{K_j} + \frac{1}{h_i} \right]^{-1}$$

and $T_{x=0}$ is the wetted surface temperature of a multilayered roof.

The effect of a wetted roof on room air temperature has been shown in Figure 5.12. One can see that there is a significant drop in room air temperature with respect to ambient air temperature.

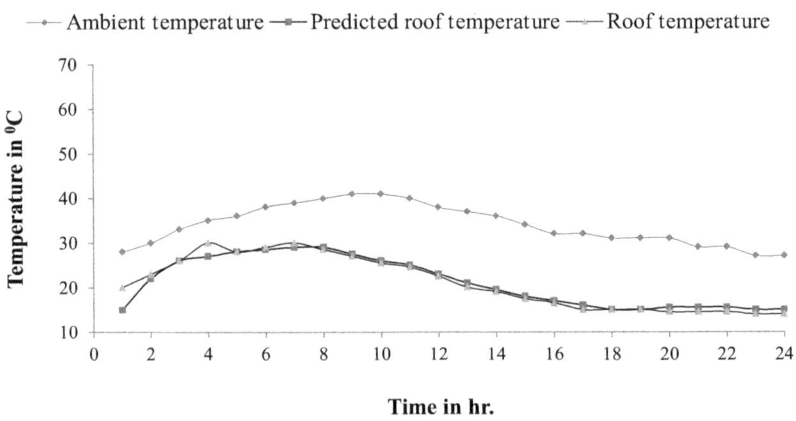

Figure 5.12: Hourly variation of room air temperature for wetted surface.

The effect of a roof pond has also been investigated by many researchers, Sodha et al. [1980a,b]. In this case, there is a possibility of water leakage and algae formation due to stagnant water.

(c) White paint/tiles

White tiles are available in the market with a high reflectivity and low thermal conductivity. These tiles are used on the outer surface of a multilayered roof to increase the reflection of solar radiation and to reduce the U value of the roof. Such studies have been carried out by many scientists, Synnefa et al. [2007]; Parker et al. (1998); Kolokotsa et al. [2012], Pisello and Cotana [2014] and David et al. (2014).

Furthermore, there are many studies carried out to reduce the solar flux coming inside an exposed roof of a room, which includes the research work of Zinzi and Agnoli [2012], Jim [2014], Tiwari et al. [1983], Kharrufa and Adil [2008], Tang and Etzion [2004] and Sodha et al. [1980a,b]. Some of the research is practical, including adding a false ceiling, and watering and plantation over the roof. The use of a false ceiling is similar to having the provision of a cabinet with an exposed thin wall. However, the roof thickness should not be compromised like a wall configuration. The roof needs to be as thick as possible to give maximum strength and resistance (minimum U value) to solar radiation. Only the watering over the roof is applied in the proposed building.

5.8 Sky irradiative cooling

The sky irradiative cooling is the long wave length radiation exchange between ambient air and the sky over the roof of a building. This is significant during night time due to low sky temperatures (T_{sky}) in comparison with the ambient air temperature (T_a) just above the roof.

Following Tiwari [2002], an expression for a radiation exchange between the ambient air and sky temperatures can be expressed as follows:

$$\dot{Q}_{r,ex} = \varepsilon\sigma[(T_a + 273)^4 - (T_{sky} + 273)^4] = \varepsilon\Delta R \tag{5.11}$$

where $\Delta R = 60\ \mathrm{W/m^2}$ is long wave length radiation exchange between the ambient air and sky temperatures and it has a constant value within the operating temperature range of the building. Here, it is to be noted that the emissivity of the roof surface (ε) should have a maximum value near 0.9 for a fast heat transfer.

Recently, Al-Obaidi et al. [2014] have reviewed the work on irradiative cooling from roof surfaces and concluded that this is responsible for minimizing the effects of outside heat penetrating into a room, as well as contributing to the reduction of internal gains through artificial lighting and electrical appliances. Irradiative cooling also reduces the load on air conditioning. A review carried out by Al-Obaidi et al. [2014] includes the work of Geetha and Velraj [2012], Givoni [1994], Santamouris et al. [2008] and Parker [2005].

6

Thermal Modelling

6.1 Experimental validation

In this section, an attempt has been made to validate the hourly variation of room air temperature on a second floor during a typical day with periodic thermal modeling as shown in Figure 6.1. An experiment has been conducted for an hourly variation of solar intensity, ambient air and room air temperatures by using a calibrated copper-constantan thermos-couple, thermometer and surya mapi (local trade name for the equipment that measures solar intensity) once a week. The hourly data was measured manually. The hourly variation of solar intensity, ambient air and room air temperatures for each floor for January 28, 2014 in winter has been shown in Figures 6.2–6.4a,b.

The design and other parameters used for thermal modeling are given in Table 6.1.

The following assumptions have been made to simplify the modeling and to write the energy balance equations for a bare roof as shown in Figure 6.1:

1. A one-dimensional transient heat conduction through the roof is considered.

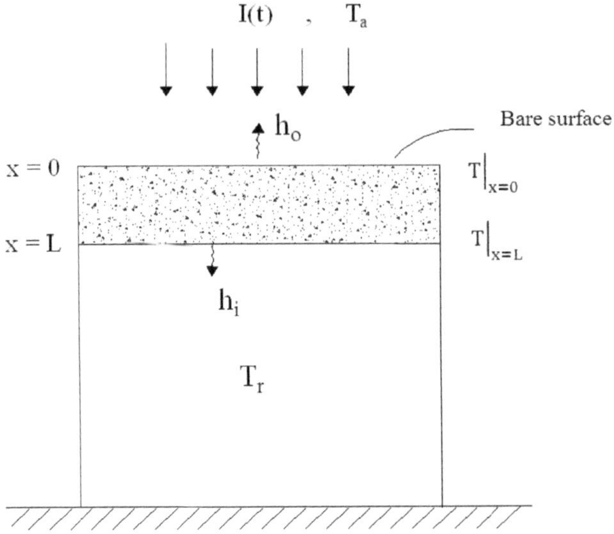

Figure 6.1: Cross-sectional view of a room in second floor for experimental validation.

2. There is no stratification in the room air temperature.
3. There are no side losses from the walls and floor except for infiltration.
4. The properties of the outside air and roof material are constant with time and temperature.
5. The mean thermal properties of the roof materials are considered due to minimum variations with value.
6. There is no infiltration between room air and ambient air.

The energy balance at different boundaries of the roof in a quasi-steady state condition is given by the following equations.

The energy balance equation for the roof at x = 0 is:

$$-k\left.\frac{\partial T}{\partial x}\right|_{x=0} = h_0\left(T_{sa} - T|_{x=0}\right) \tag{6.1}$$

where T_{sa} is a solair temperature same as given by Eq. (5.2b).

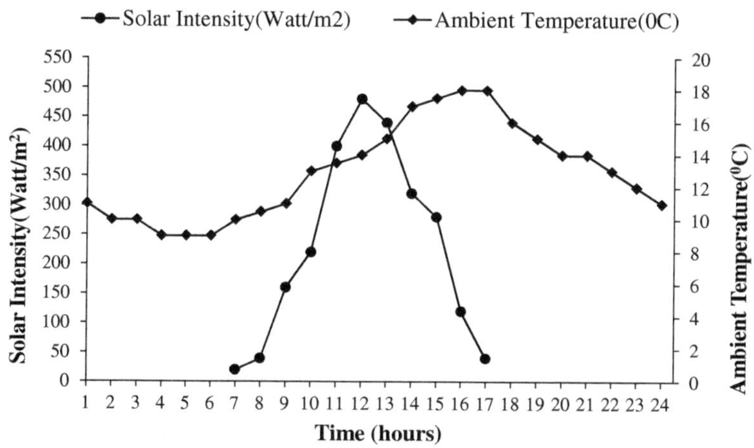

Figure 6.2: Hourly variation of solar intensity I(t) and ambient temperature T_a for a typical day of January 28, 2014, Varanasi, India.

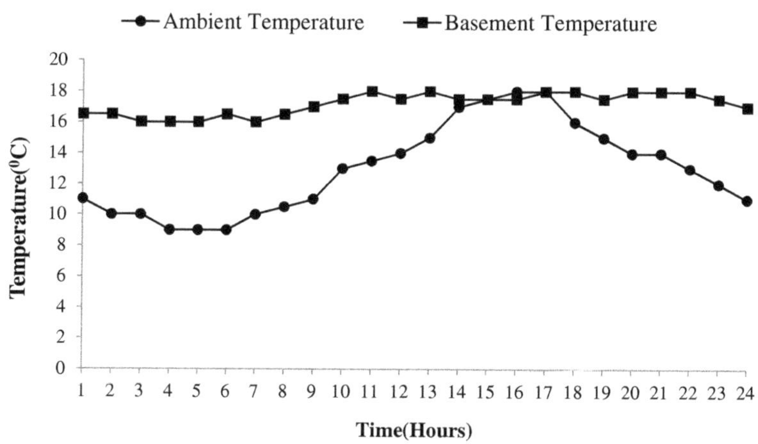

Figure 6.3a: Hourly variations of ambient temperature and basement air temperature.

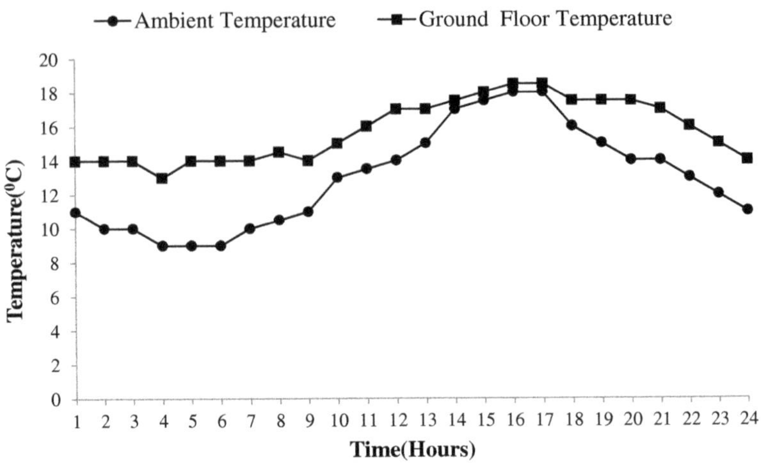

Figure 6.3b: Hourly variations of ambient air temperature and ground floor air temperature.

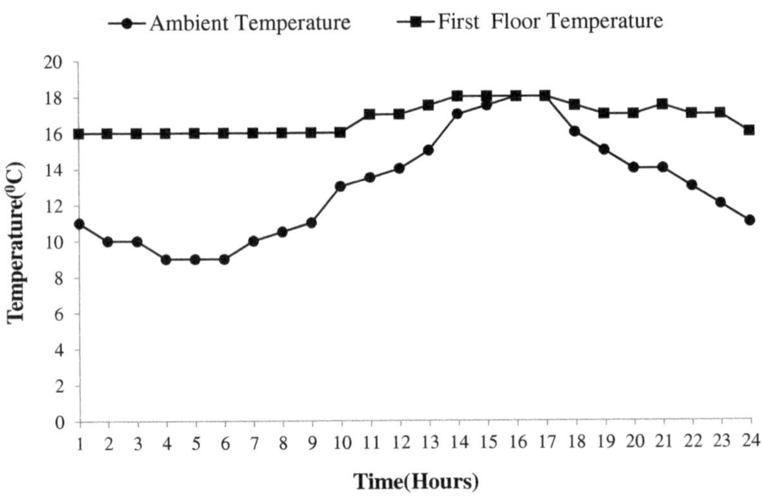

Figure 6.4a: Hourly variations of ambient air temperature and first floor air temperature.

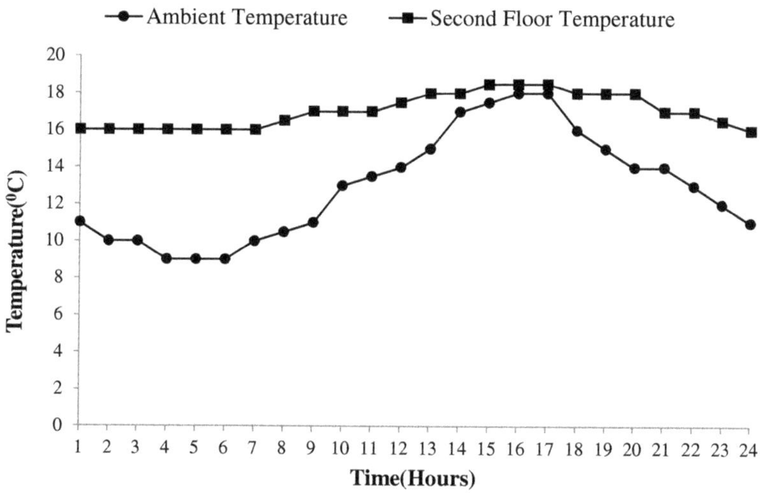

Figure 6.4b: Hourly variations of ambient air temperature and second floor air temperature.

Table 6.1: Design parameters of SBC for thermal modeling.

Parameters	Values	Parameters	Values
ρ_a	1.2 kg m^{-3}	h_o	5.7 W/m^2 °C
C_a	1005 J kg^{-1}K^{-1}	h_i	2.8 W/m^2 °C
C	1130 J kg^{-1}K^{-1}	h_c	2.8 W/m^2 °C
ΔR	60 W/m^2	$\propto$	0.4
k_g	0.816 W/m °C	$l_r \times b_r \times h$	3 m × 3 m × 3 m
ρ	2400 kg m^{-3}	A_r	9 m^2
k	1.296 W/m °C	V	27 m^3
ε	0.9		

It can be evaluated for a given I(t), T_a as shown in Figure 6.2 and numerical data given in Table 6.1.

Referring to Figure 6.1, the energy balance equation for the roof at x = L is written as follows:

$$-k \left.\frac{\partial T}{\partial x}\right]_{x=L} = h_i \left(T|_{x=L} - T_r\right) \tag{6.2}$$

The available useful thermal energy is used to heat the room. The energy balance for the space heating of the room is given by:

$$M_a C_a \frac{\partial T_r}{\partial t} = h_i A_r (T|_{x=L} - T_r) - 0.33 N_2 V_2 (T_r - T_a) \tag{6.3}$$

Since the solar radiation and ambient air temperature are periodic in nature as shown in Figure 6.2, the solair temperature (T_{sa}, Eq. 5.9), ambient air temperature and room air temperature are also represented mathematically in the form of a Fourier series. These are as follows:

$$T_{sa} = T_{sa0} + Re \sum_{n=1}^{6} T_{san} e^{i(n\omega t - \phi_n)} \tag{6.4a}$$

$$T_a = T_{a0} + Re \sum_{n=1}^{6} T_{an} e^{i(n\omega t - \gamma_n)} \tag{6.4b}$$

$$T_r = T_{r0} + Re \sum_{n=1}^{6} T_{rn} e^{in\omega t} \tag{6.5}$$

The Fourier coefficients of solair and ambient air temperatures [Eqs. (6.4a and 6.4b)] for January 28, 2014 are given in Table 6.2.

Equations (6.1)–(6.3) can form a matrix of 3 × 3 for time independent and dependent part which can be solved by a matrix inversion by using Matlab.

The results obtained from the thermal model as described in this study have been validated with second floor room air temperatures at SBC. The total roof thickness for the second floor is approximately 0.40 m. The theoretical room air temperature for a zero number of air

Table 6.2: The Fourier coefficient of: (a) total solar intensity falling on the roof and (b) ambient air temperature.

n	0	1	2	3	4	5	6
$(a)\ T_{an}$	13.4250	−4.0891	0.4801	0.0112	0.0190	−0.0209	−0.0135
γ_n		52.6580	67.1889	0.0000	−19.1066	9.7869	−45.0000
$(b)\ T_{san}$	35.1833	−37.9016	15.3278	−2.5782	−0.4226	0.8149	−0.9947
ϕ_n		7.9552	3.9330	−1.3379	0.8923	−4.1028	−4.9523

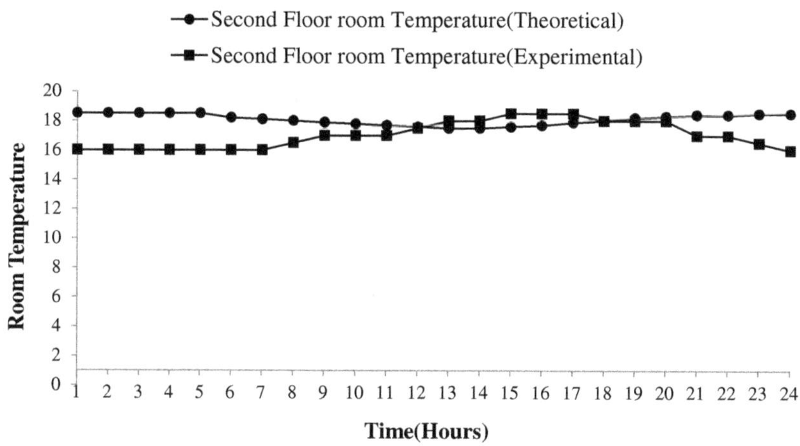

Figure 6.5: Hourly variation of room air temperature for bare roof (0.40m roof thickness) and on the second floor of SODHA BERS COMPLEX.

changes is in agreement with experimental value as shown in Figure 6.7. Due to cold climatic conditions in January it is assumed that the doors and windows will be tightly sealed so that minimum air infiltration takes place. Figure 6.5 shows the hourly variation of theoretical and experimental result of room air temperature of the second floor. There is a small variation between theoretical and experimental observations during the night due to following reasons:

(a) a higher U value through the east and west walls and clerestory window in south wall; and

(b) infiltration due to the opening of a sliding door provided in the south wall.

Furthermore, the correlation coefficient between theoretical and experimental observations is found to be about 0.96 and a standard deviation of 4.5%. Therefore there is a fair agreement between theoretical and experimental observations.

The results for the hourly room air temperatures on each floor for January 28, 2014 in winter, as shown in Figures 6.3a,b and 6.4a,b will be discussed in Section 9.

6.2 Monthly Thermal Energy Saving

The monthly average room air temperature of each floor has been evaluated by averaging data for each hour during each month. The results for each month for the basement and second floor have been shown in Figures 6.6 and 6.7. It is seen from Figure 6.6 that the average room air temperature during the harsh summer months of May and June for the basement (earth shelter) is 7 – 8.65°C lower than the ambient air temperature. This indicates that there is passive cooling available due to ground thermal storage. Furthermore, it is also observed that there is an additional cooling effect in the room near the rain harvesting in the basement on the north side. The average room air temperature of the ground floor during each month is approximately same as the ambient air temperature due to the openness of this floor, as shown in Figure 6.6. The rise of 2 – 3°C on the first floor during harsh the winter months (December and January) shows that there is passive heating due to a clerestory window in the east portion, as seen from Figure 6.7. The same figure also shows that the results are on expected line for second floor. The room air temperature on the second floor (see Figure 6.7) is less than the ambient temperature in summer and more

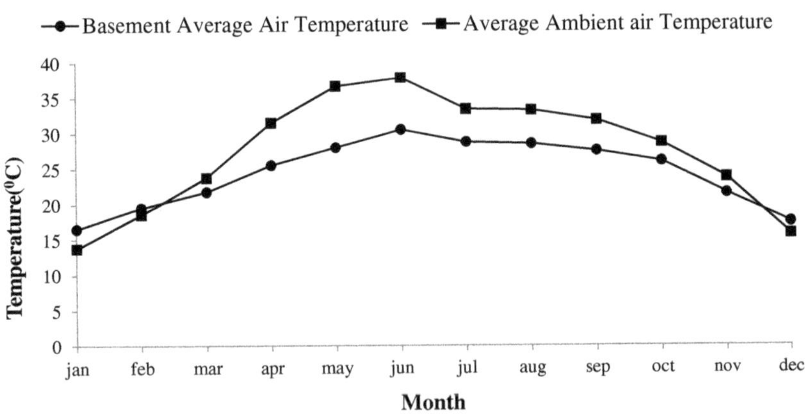

Figure 6.6: Monthly variations of average ambient temperature and average basement air temperature.

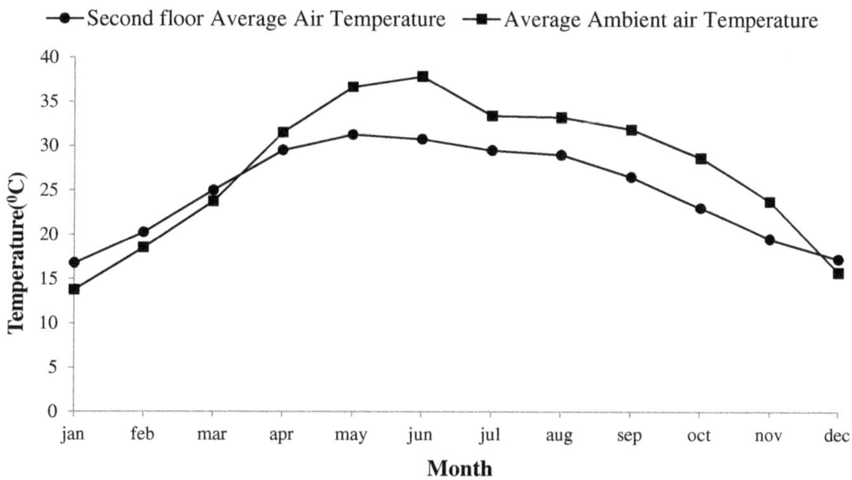

Figure 6.7: Monthly variations of average ambient temperature and average second floor temperature.

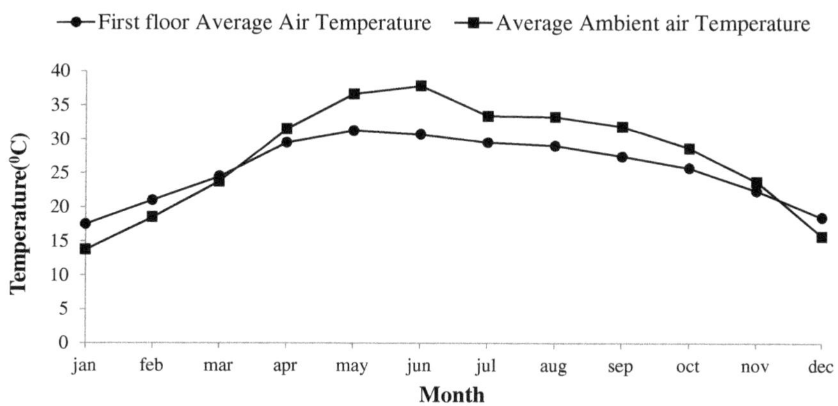

Figure 6.8: Monthly variations of average ambient air temperature and average first floor air temperature.

in winter. During the winter months, the second floor is more suitable and comfortable than the ground floor. The results for the first floor are similar to that of the second floor.

Based on the data of the room air temperature, the monthly energy saving has been evaluated for each floor as follows:
The total covered area of SBC = 152.42 m^2
The total volume of air contained on one floor in SBC = (152.42×3.6) m^3 = 548.71 m^3
The mass of the air on one floor (M_a) = (548.71×1.19)kg = 652.96 kg
An average temperature difference between the room air and ambient air temperature of $\Delta T = 6°C$ is then considered for each floor.
The hourly heat gain by one floor at SBC (Q)

$$= M_{a.}C_{a.}\Delta T = (652.96 \times 1 \times 6)\,\text{kJ} = 3917.76\,\text{kJ}$$
$$= 3917.76 \times 2.778 \times 10^{-4} = 1.08835\,\text{kWh}$$

where $C_a = 1\,\text{kJ/kg/°C}$ is the specific heat of the air.
The daily heat gain for one floor = $1.088 \times 24\,\text{kWh} = 26.112\,\text{kWh}$
The annual heat gain for one floor = $26.112 \times 365\,\text{kWh} = 9530.88\,\text{kWh}$
There are three residential floors in SBC.
Therefore the total annual thermal heat gain for the complete building

$$= 9530.88 \times 3\,\text{kWh}$$
$$= 28592.64\,\text{kWh} \tag{6.6}$$

Similarly, the total annual thermal heat gain for the complete building can also be evaluated for different $\Delta T = 4°C$, $8°C$ and $10°C$ respectively.

6.3 Energy saving due to natural day lighting

Natural day lighting in the building refers to the illumination (Lux) inside each room through (a) clerestory/wall windows (facade) (Figures 5.5a and 5.5b), wind towers (Figures 5.6a,b and 5.7) during sunshine hours/day. It is an important issue and essential component of the solar passive building for the architectures [Roche et al., 2000, Chow et al., 2013, Goiaa et al., 2013]. This provides illumination without a heat source unlike artificial lighting and saves on conventional (fossil) fuels. Additionally, day lighting is an impressive and sustainable development strategy for buildings due to the increased visual comfort, physical health, energy-efficiency and green building developments (Li, 2007; Edwards and Torcellini, 2002, Chan and Tzempelikos, 2013). For an attractive indoor environment, day lighting inside a room can be provided by appropriate glass windows (faced) that open [Sherif et al., 2012, Zheng et al., 2013]. The provision of the proper day lighting in any single or multi-story building consume less energy and provides a good visual performance [Wright and Letherman, 1998, CIBSE, 1999, Calcagni and Paroncini, 2004, Nabil and Mardaljevic, 2006]. Many other authors have studied natural day lighting for different shape/story buildings under different climatic conditions (Li and Lam [2001], Wong et al. [2012], Mahlia et al. [2011], Sudan and Tiwari [2014], Gago et al. [2015]). These authors have observed that there is a significant energy saving due to illumination by natural day lighting. Their work includes annual energy saving, exergy analysis, CO_2 mitigation and modeling etc. In view of this, there is a need to study day lighting for a known Indian climatic condition.

For a single story building, illumination can be achieved through the roof as shown in Figure 5.3. In SBC, the concept of clerestory/wall windows (façade) and wind towers has been used (Figures 5.5a and 5.6a). As shown in Figures 5.5b and 5.6b, due to the provision of a glazing area in the east side of the north portion of the building and lower partition walls up to height of 2.4 m × 0.11 m, there is no need for artificial light throughout the year. On the basis of experimental studies, Sudan and Tiwari [2014], it has been observed that the annual energy saving from diffuse radiation is 12.8 kWh/m^2 at SBC. There is a total area of 152.42 m^2 illuminated with diffuse radiation in SBC, and the annual energy saving from natural day lighting for three floors can be evaluated as follows:

Annual energy saving due to day lighting E_{aout} = Annual energy saving by diffuse radiation × area × no. of floors:

$$= 12.8 \times 152.42 \times 3 = 5852.93\,\text{kWh} \tag{6.7}$$

6.4 An overall energy saving at SBC

An overall energy saving including thermal and day lighting for SBC can be written as follows:

$$\begin{aligned}
&\text{An overall energy saving for SBC}\\
&\quad = \text{Energy saving by thermal heat gain/loss (Eq. 6.6)}\\
&\qquad + \text{Energy saving by day lighting (Eq. 6.7)}\\
&\quad = 28592.64\,\text{kWh} + 5852.93\,\text{kWh} = 34,445.568\,\text{kWh} \qquad (6.8)
\end{aligned}$$

Similarly, an overall energy saving for the complete building can also be evaluated for different $\Delta\text{T} = 4°\text{C}$, 8 °C and 10 °C respectively. The results are given in Table 8.4.

7

Cost of Construction (Initial Investment)

The breakdown costs of each material used for the construction of the building is given in Table 7.1. It can be seen that maximum cost incurred during the construction of the building is labor and reinforcement materials.

The initial cost of the building (P_I) in rupees is € 4818710 without the cost of the land. The market value of the land is more than the cost of the building at the time of construction in Varanasi (UP).

Table 7.1: Cost of materials used for SBC (1$ = ₹60).

S.No.	Name of material/ work	Costing (₹)	Costing ($)	S.No.	Name of material/ work	Costing (₹)	Costing ($)
1	Reinforcement	927708	15461.8	8	Glass work	94077	1567.95
2	Stone ballast	232268	3871.13	9	Lime/putty/pop	59070	984.5
3	Sand	260972	4349.53	10	Painting work	93893	1564.88
4	Bricks	203153	3385.88	11	Electrification	193136	3218.93
5	Iron work	304996	5083.27	12	Sanitary work	279397	4656.62
6	Wood work	83424	1390.4	13	Cement	733500	12225
7	PVC work	97402	1623.37	14	Labor with T&P	1255714	20928.6
					Initial Investment (P_I)	4818710	80311.83

8

CO_2 Analysis

In order to evaluate the carbon credit, the embodied energy of SBC should be known. So we will evaluate the embodied energy of SBC.

8.1 Embodied energy

In order to study the energy analysis, it is important to know the mass of each material ($\mathbf{m_i}$) used for the construction of the building and its energy density (e_i) to evaluate the embodied energy, as shown in Tables 5a and 5b respectively:

By using the data from Tables 8.1a and 8.1b, the embodied energy can be evaluated as follows:

$$E_{in} = \sum_{i}^{n} m_i e_i \tag{8.1}$$

The embodied energy of materials used for SBC has been evaluated by using Eq. 8.1 and is shown in Table 8.1c.

Furthermore, the embodied energy ($\mathrm{E_{in}}$) can be classified into four different conditions, namely:

Case (A): Existing SBC

Table 8.1a: Breakdown of materials (m_i) used for SBC.

Serial	Name of Material (m_i)	Qty (kg)	Serial	Name of material	Qty (kg)
1	Cement	122250	10	Wood	8200
2	Bricks/tiles	337935	11	PVC/plastic	168
3	Stone ballast	447590	12	Paint (oil bound)	210
4	Sand	540937	13	Paint (water bound)	1125
5	Lime/pop/putty	2220	14	Marble/makrana stone	525
6	Reinforcement	22574	15	Ceramic tiles/ sanitary ware	4457
7	Aluminum	34	16	Iron	4458
8	Glass	1248	17	Copper	180
9	Plywood	10950	18	Brass	150

Table 8.1b: Energy density of each material (e_i) used for the construction of SBC, Wikipedia.

Serial	Material/Name of work (e_i)	Energy Density (MJ/Kg)
1	Sand	0.1
2	Stone ballast	0.12
3	Marble/makrana stone	2
4	Bricks/tiles	2.5
5	Lime/pop/putty	5.3
6	Cement	5.5
7	Wood	8.5
8	Ceramic tiles/sanitary ware	10
9	Plywood	15
10	Glass	18
11	Iron	25
12	Reinforcement	35
13	Paint (water bound)	53
14	Brass	62
15	Paint (oil bound)	68
16	Copper	70.6
17	PVC/plastic	77
18	Aluminium	155

Case (B): Without activities/process

Case (C): Without activities/process and transportation

Case (D): Case (C) with recycled material.

Table 8.1c: The embodied energy (E_{in}) of SBC.

Material/Name of work	Embodied energy (MJ)	Embodied energy (kWh)	Embodied Energy/Year Lifespan 100 Years (kWh)
Cement	672375	186770.83	1867.71
Bricks/tiles	844837.5	234677.08	2346.77
Stone ballast	53710.8	14919.67	149.20
Sand	43815.897	12171.08	121.71
Lime/pop/putty	11766	3268.33	32.68
Reinforcement	799119.6	221977.67	2219.78
Aluminum	5270	1463.89	14.64
Glass	18720	5200.00	52.00
Plywood	164250	45625.00	456.25
Wood	69700	19361.11	193.61
PVC/plastic	12936	3593.33	35.93
Paint (oil bound)	14280	3966.67	39.67
Paint (water bound)	59625	16562.50	165.63
Marble/makrana stone	1050	291.67	2.92
Ceramic tiles/sanitary ware	89140	24761.11	247.61
Iron	111450	30958.33	309.58
Copper	7560	2100.00	21.00
Brass	6600	1833.33	18.33
Activities/process	867664	241017.78	2410.18
Transportation	180844	50234.44	502.34
Total	4034713.80	1120753.83	11207.54

The results of the embodied energy for the different cases have been shown in Table 8.2 which will be used for energy analysis.

From Table 8.2 one can observe that the existing SBC has a maximum embodied energy, while in the case without activities/process and transportation with recycled material it has a minimum embodied energy per year. Therefore, the life of building will also be much less when recycled materials have been used for construction.

8.2 CO_2 Emission

Figure 8.1 shows the CO_2 emission by each individual component of building. It is seen that the total CO_2 emission is 139.38 tons per year.

Table 8.2: Embodied energy of materials used for SBC each year yearly by 100 year lifespan.

Cases	Embodied energy/year (kWh/Year)
A	11207.54
B	8797.36
C	8295.02
D	8131.98

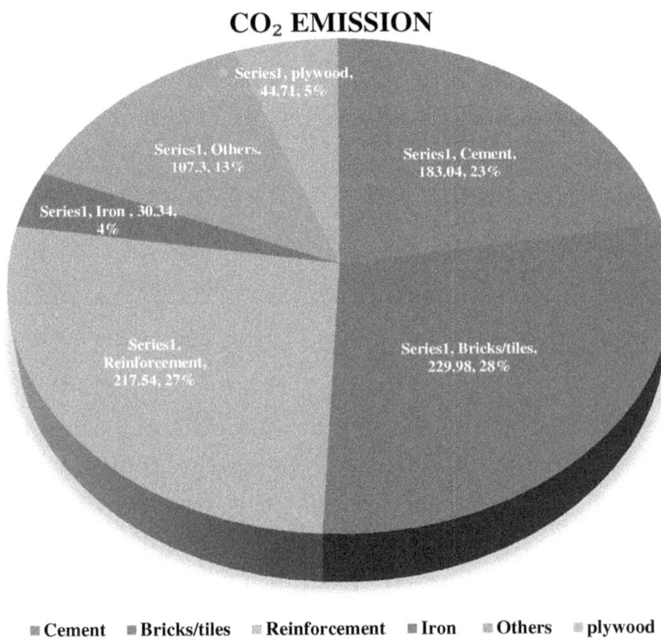

Figure 8.1: CO_2 emissions due to each material.

The major contributors of CO_2 emission are brick, cement and reinforcements (78%). It is to be noted that brick contributes about 28% to this total due to the use of a 0.115 m brick exterior and partition wall, otherwise its contribution would have been even more (for a thicker wall).

If one unit (kWh) of conventional (fossil fuel) power is used by a consumer in a residential house, and L_a (0.20) is the losses in fraction due to poor domestic appliances, then the available transmitted power should be $\frac{1}{1-L_a}$ units (kWh).

If L_{td} (0.40) is the transmission and distribution losses in fraction from power plant site, then the power that needs to be generated at power plant site becomes:

$$P_{site} = \frac{1}{1-L_a} \times \frac{1}{1-L_{td}} \text{kWh (units)}$$

The average CO_2 equivalent intensity for the electricity generation from the best coal quality is approximately 0.98 kg of CO_2 per kWh at the source. Thus, the amount of CO_2 emission is expressed as:

$$[CO_2]_{emission} = \frac{1}{1-L_a} \times \frac{1}{1-L_{td}} \times 0.98\,\text{kg per kWh power consumption by the consumer.}$$

The annual CO_2 emission, Tiwari and Mishra [2011] can be expressed as:

$$[CO_2]_{emission} \text{ per year} = \frac{E_{in}}{L} \times \frac{1}{1-L_a} \times \frac{1}{1-L_{td}} \times 0.98\,\text{kg} \quad (8.2)$$

where E_{in} is the embodied energy (kWh) per year as given in Table 8.2 for the life of a building of $L = 100$ years.

The annual CO_2 emission per year for case A of Table 8.2

$$= 11207.54 \times \frac{1}{1-0.20} \times \frac{1}{1-0.4} \times 0.98\,\text{kg}$$

$$= 22885.79 \text{ kg } = 22.87 \text{ tons} \quad (8.3)$$

Furthermore, the CO_2 emission per year is given in Table 8.3 for the following cases:

Case (A): Existing SBC

Case (B): Without activities/process

Table 8.3: CO_2 Emission for L = 100 years.

Cases	CO_2 Emission (kg)/Year
A	22885.79
B	17964.21
C	16938.42
D	13609.26

Case (C): Without activities/process and transportation

Case (D): Case (C) with recycled material.

It can be noted from Table 8.3 that the CO_2 emission is minimal for recycled reinforcement, case D. In this instance, the effective lifespan of the building may be reduced, and in a future situation it could become a danger to life due to a possible building collapse.

8.3 CO_2 mitigation of SBC

The net CO_2 mitigation over the lifetime of the system, n years, Agrawal and Tiwari [2010] is given by:

Total CO_2 mitigation – total CO_2 emission

$$= (E_{aout} \times L - E_{in}) \times \frac{1}{1 - L_a} \times \frac{1}{1 - L_{td}} \times 0.98 \times 10^{-3} \text{ tons} \quad (8.4)$$

The performance of SBC has been studied on the basis of a monthly average room temperature over a complete one year period. Based on the monthly performance of SBC building, the CO_2 mitigation and energy payback period has been evaluated on the basis of energy analysis.

After substituting an appropriate value $E_{aout} = 34,445.568$ kWh (Eq. 6.8) and $E_{in} = 11207.54$ per year for case A (Table 8.2), net CO_2

mitigation for 100 years, for case A is obtained by (Eq. 8.4) as:

$$= (34,445.568 \times 100 - 1120754) \times \frac{1}{1-0.20} \times \frac{1}{1-0.40} \times 0.98 \times 10^{-3}$$
$$= 4740.56\, tons \tag{8.5}$$

For a lower value of embodied energy e.g. recycled materials namely E_{in} = 8131.98 kWh per year for case D (Table 8.2) and for a longer life of the building at 300 years, the net CO_2 mitigation for case D is given by:

$$(34,445.568 \times 300 - 813198) \times \frac{1}{1-0.20} \times \frac{1}{1-0.40} \times 0.98 \times 10^{-3}$$
$$= 9520.47 \text{ tons}$$

which is much higher than in case A.

8.4 Energy payback time

The energy payback time of SBC can be evaluated by using the following expression:

$$\text{Energy Payback Time (EPBT)} = \frac{\text{Embodied energy, Table 8.2}}{\text{an overall energy saving, Eq. 6.8}} \tag{8.6}$$

Table 8.4: The energy payback time of SBC.

Average temperature difference (ΔT) between room air and ambient air temperature (°C)	The total annual thermal energy saving for SBC (E_{aout}) (kWh)	The energy payback time (years)
4	24932.208	44.95
6	34445.568	32.54
8	43985.208	25.48
10	53524.848	20.94

The above calculated energy payback time does not consider the land cost. Here it is important to mention that the building has been in use for 365 days a year.

8.5 Carbon credit earned by SBC

If the CO_2 emission is being traded at 10 US$ per ton of CO_2 mitigation, then the carbon credit earned by the system is obtained as:

$$\text{The carbon credit earned} = 4740.56 \times 10 = \text{US\$ } 47405.6 \qquad (8.7)$$

The above calculated carbon credit will have a higher value for an increased value of CO_2 mitigation.

9

Lifecycle Cost Analysis

In order to evaluate the energy cost saving per kWh, we have made the following assumptions:

(i) Expected life of the building (n) = 100 years

(ii) Operational and maintenance cost per year = 0.5% of initial cost $= 0.005 \times P_I$

(iii) The salvage value of SBC has been considered as 0.10, 0.15, 0.20, 0.25 and 0.30 of the future value of the initial investment.

(iv) The rate of interest (i) = 5%

The cash flow diagram of SBC is shown in Figure 9.1.

The salvage value of the building after 100 years has been considered as:

$$S = F \times P_I \times (1 + i)^n \tag{9.1}$$

where F is the fractional salvage, which is always less than one. In the present study, the salvage value has been considered as 0.10, 0.15, 0.20, 0.25 and 0.30 fraction of the present investment (P_I).

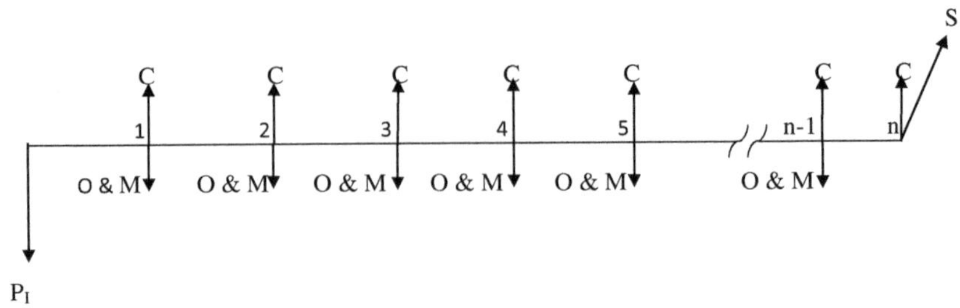

Figure 9.1: Cash flow diagram of SODHA BERS COMPLEX (SBC).

Following Tiwari (2012), the uniform annual cost (Unacost) of the cash flow diagram of Figure 9.1 can be written as follows:

$$C_A = \left[P_I \times \frac{i \times (1+i)^n}{(1+i)^n - 1} + O\&M - S \times \frac{i}{(1+i)^n - 1} - C\right] \tag{9.2}$$

Where:
Initial investment (P_I) = ₹4818710.00 (from Table 7.1)
Operational and maintenance cost (O&M) = $0.005 \times P_I$ = ₹24094.00

$$\begin{aligned} \text{Carbon credit earned per year (C)} &= \frac{\text{US\$47405.6}}{100} = \text{US\$ 474.056} \\ &= ₹474.056 \times 60 \\ &= ₹28443.36 \text{ (from Eq. 8.7)} \end{aligned}$$

The unit cost per kWh of energy saved for SBC can be written as follows:

$$\text{Unit cost per kWh} = \frac{\text{Uniform annualized cost } (C_A)}{\text{Annual energy saving } (E_a)} \tag{9.3}$$

The values of the uniform annual cost (C_A) and unit cost/kWh for different salvage value have been calculated by using Eqs. 9.2 and 9.3 respectively. The results can be seen in Tables 9.1–9.4.

Table 9.1: Effect of salvage value on uniform annualized cost (C_A) and unit cost/kWh for $\Delta T = 4°C$.

Case	Salvage value (₹)	Uniform annualized cost C_A (₹)	Annual energy saving (kWh) (from Table 8.4)	Unit Cost/kWh (₹)
(a)	63366642.62 (for F = 0.10)	214154.56	24932.21	8.59
(b)	95049963.93 (for F = 0.15)	202015.47	24932.21	8.10
(c)	126733285.24 (for F = 0.20)	189876.38	24932.21	7.62
(d)	158416606.55 (for F = 0.25)	177737.30	24932.21	7.13
(e)	190099927.86 (for F = 0.30)	165598.22	24932.21	6.64

Table 9.2: Effect of salvage value on uniform annualized cost (C_A) and unit cost/kWh for $\Delta T = 6°$ C.

Case	Salvage value (₹)	Uniform annualized cost C_A (₹)	Annual energy saving (kWh) (from Table 8.4)	Unit Cost/kWh (₹)
(a)	63366642.62 (for F = 0.10)	214154.56	34445.57	6.22
(b)	95049963.93 (for F = 0.15)	202015.47	34445.57	5.86
(c)	126733285.24 (for F = 0.20)	189876.38	34445.57	5.51
(d)	158416606.55 (for F = 0.25)	177737.30	34445.57	5.16
(e)	190099927.86 (for F = 0.30)	165598.22	34445.57	4.81

Table 9.3: Effect of salvage value on uniform annualized cost (C_A) and unit cost/kWh for $\Delta T = 8°C$.

Case	Salvage value (₹)	Uniform annualized cost C_A (₹)	Annual energy saving (kWh) (from Table 8.4)	Unit Cost/kWh (₹)
(a)	63366642.62 (for F = 0.10)	214154.56	43985.21	4.87
(b)	95049963.93 (for F = 0.15)	202015.47	43985.21	4.59
(c)	126733285.24 (for F = 0.20)	189876.38	43985.21	4.32
(d)	158416606.55 (for F = 0.25)	177737.30	43985.21	4.04
(e)	190099927.86 (for F = 0.30)	165598.22	43985.21	3.76

Table 9.4: Effect of salvage value on uniform annualized cost (C_A) and unit cost/kWh for $\Delta T = 10°C$.

Case	Salvage value (₹)	Uniform annualized cost C_A (₹)	Annual energy saving (kWh) (from Table 8.4)	Unit Cost/kWh (₹)
(a)	63366642.62 (for F = 0.10)	214154.56	53524.85	4.00
(b)	95049963.93 (for F = 0.15)	202015.47	53524.85	3.77
(c)	126733285.24 (for F = 0.20)	189876.38	53524.85	3.55
(d)	158416606.55 (for F = 0.25)	177737.30	53524.85	3.32
(e)	190099927.86 (for F = 0.30)	165598.22	53524.85	3.09

10

Result and Discussion

The hourly variation of the ambient and room air temperatures of the basement (earth sheltering) at SBC has been shown in Figure 6.3a. It can be observed that these variants are at a maximum during the months of May and June respectively. The average monthly temperature difference between the air temperatures of the basement and ambient air indicates that there is passive heating/cooling in the basement during the winter and summer months respectively. Passive heating and passive cooling occurs because of heat gain and loss in winter and summer between the room air of the basement and surrounding walls that are attached to heavy structure earth. The maximum basement room air temperature is about 30.5 °C in June which can be comfortable from a rural point of view against ambient temperature of 37.85°C (Figure 6.6).

The variation of the monthly room air temperature for the first and second floors is approximately same as shown in Figure 6.7. The monthly average temperature difference between the first and second floor with ambient air indicates that in winter months (December to April) there is passive heating, which is for two months more than in the basement. It also shows that during the other months there is passive cooling. However, it can also be noted that there is a maximum temperature difference of 7.1°C during June, which provides a comfortable living in the rooms of the first and the second floors. It is also seen that we have a good heating and cooling effect in the case of first and second floors due to better heat transfer from WT-II and walls.

Table 8.4 shows that energy payback period decreases with the increase in temperature difference as expected. The energy payback period of 44.95 years is very high against the life of 100 years of the building. This result is evaluated on the initial stage of building and can

Figure 10.1: Proposed building integrated photovoltaic thermal roof.

be further optimized by having more passive cooling/heating concepts namely:

(i) Plastering the outside of the building with waterproof white paint (cooling)

(ii) Proper opening/closing of the clerestory window and having wind towers provided on each floor (cooling/heating)

(iii) Having an air cavity cabinet on each floor to reduce futher U-value

(iv) Constructing a building integrated photoboltaic thermal (BiPVT) roof system, see Figure 10.1.

Tables 9.1–9.4 give the information on the unit cost of energy saving under different conditions. It should be noted that the salvage value and temperature difference both play an important role. It is seen that the unit cost of energy saving varies between ₹3–₹9.

11

Recommendations

For a sustainable environment and climate on the basis of present case studies, the following recommendations for the passive cooling of a building have been made:

- The exposed area of a south-facing building in the northern hemisphere should be minimum
- The minimum height of a room on each floor in a multi-story building should be 3.6 m
- The exposed area of each wall should have a minimum glazing area (Figure 4.2)
- An air cavity (cabinet) integrated into an unglazed Trombe wall with storage should be considered for urban housing (Figure 4.1)
- A basement (earth sheltering) should be made for community service
- Cross-ventilation in urban housing complexes should be made with wind towers if possible

- Green roofs should be encouraged for passive cooling
- To make the building self-sustainable and grid independent, the provision of a Building Integrated Photovoltaic Thermal system should be incorporated as shown below in Figure 10.1.

Figure 11.1: Photograph of inauguration ceremony of SODHA BERS COMPLEX on February 8, 2012.

Acknowledgements

The authors are thankful to Prof. Brian Norton, President, DIT, Ireland; Prof. Christophe Menezo, Germany; Prof. Wolfram Sparber, Italy; Prof. TT Chow, Hong Kong; and Prof. B.K. Bala, Bangladesh, for their presence as guests of honor during the inauguration (see Figure 11.1) of SODHA BERS COMPLEX, Varanasi, on February 8, 2012. We are also grateful to both reviewers for their constructive suggestions to improve the quality of the case study. The authors are also grateful to the Department of Science and Technology (DST) for their partial financial support to carry out the review and case study on passive cooling.

References

Acred, A. and G. R. Hunt. Stack ventilation in multi-storey atrium buildings: A dimensionless design approach. *Building and Environment*, 72:44–52, 2014.

Agrawal, B. and G. Tiwari. Building integrated photovoltaic thermal systems. *Royal Society of Chemistry* (RSC), (UK), 2010.

Al-Obaidi, K. M., I. Mazran, and M. A. Rahman. Passive cooling techniques through reflective and radiative roofs in tropical houses in southeast asia: A literature review. *Frontiers of Architectural Research*, 3(3):283–297, 2014.

Al-Sanea, S. A. and M. Zedan. Improving thermal performance of building walls by optimizing insulation layer distribution and thickness for same thermal mass. *Applied Energy*, 88(9):3113–3124, 2011.

Antar, M. A. Thermal radiation role in conjugate heat transfer across a multiple cavity building block. *Energy*, 35(8):3508–3516, 2010.

Bansal, N. K. and G. Minke. *Climatic Zones and Rural Housing in India.* Germany, 1988.

Bisoniya, T. S., K. Anil, and B. Prashant. Experimental and analytical studies of earth–air heat exchanger (EAHE) systems in india: A review. *Renewable and Sustainable Energy Reviews*, 19:238–246, 2013.

Bojic, M., K. Johannes, and F. Kuznik. Optimizing energy and environmental performance of passive Trombe wall. *Energy and Buildings*, 70:279–286, 2014.

Calcagni, B. and M. Paroncini. Daylight factor prediction in atria building designs. *Solar Energy*, 76:669–682, 2004.

Chan, Y. C. and A. Tzempelikos. Efficient venetian blind control strategies considering daylight utilization and glare protection. *Solar Energy*, 98: 241–254, 2013.

Charde, M. and R. Gupta. Design development and thermal performance evaluation of static sunshade and brick cavity wall: An experimental study. *Energy and Buildings*, 60:210–216, 2013.

Chen, X., H. Yang, and L. Lu. A comprehensive review on passive design approaches in green building rating tools. *Renewable and Sustainable Energy Reviews*, 50:1425–1436, 2015.

Chow, S. K. H., D. H. W. Li, E. W. M. Lee, and J. C. Lam. Analysis and prediction of daylighting and energy performance in atrium spaces using daylight-linked lighting controls. *Applied Energy*, 112:1016–1024, 2013.

Chu, C.-R. and C. Bo-Fan. Wind-driven cross-ventilation in long buildings. *Building and Environment*, 80:150–158, 2014.

CIBSE. Daylighting and window design. *Lighting Guide LG10*, 1999.

Deo, A. Energy conservation in building. M.Tech. Thesis, CES, IIT Delhi, 2013.

Deo, A., G. Agam, A. Tiwari, and G. N. Tiwari. Rooftop electrical power generation for rural applications. Communicated, Energy Policy, 2015.

Edwards, L. and P. Torcellini. A literature review of the effects of natural light on building occupants. NREL/TP-550-30769, 2002.

Faggianelli, G. A., A. Brun, E. Wurtz, and M. Muselli. Natural crossventilation in buildings on mediterranean coastal zones. *Energy and Buildings*, 77:206–218, 2014.

Gago, E. J., T. Muneer, M. Knez, and H. Köster. Natural light controls and guides in buildings. Energy saving for electrical lighting, reduction of cooling load. *Renewable and Sustainable Energy Reviews*, 41:1–13, 2015.

Gan, G. A parametric study of Trombe wall for passive cooling of buildings. *Energy and Buildings*, 27:37–43, 1998.

Geetha, N. B. and R. Velraj. Passive cooling methods for energy efficient buildings with and without thermal energy storage — a review. *Energy Edu. Sci. Technol. PartA: Energy Sci.*, 29(2):913–946, 2012.

Givoni, B. *Passive Low Energy Cooling of Buildings*. John Wiley & Sons. Vanm Nostrand Reinhold Co, New York, 1994.

Goiaa, F., M. Haase, and M. Perinoa. Optimizing the configuration of a facade module for office buildings by means of integrated thermal and lighting simulations in a total energy perspective. *Applied Energy*, 108:515–27, 2013.

Haw, L. C., S. Omidreza, M. Y. Sulaiman, M. Sohif, and S. Kamaruzzaman. Empirical study of a wind-induced natural ventilation tower under hot and humid climatic conditions. *Energy and Buildings*, 52:28–38, 2012.

Jie, J., Y. Hua, P. Gang, and L. Jianping. Study of PV- trombe wall installed in a fenestrated room with heat storage. *Applied Thermal Engineering*, 27: 1507–1515, 2007.

Jim, C. Y. Building thermal-insulation effect on ambient and indoor thermal performance of green roofs. *Ecological Engineering*, 69:265–275, 2014.

Kharrufa, S. N. and Y. Adil. Roof pond cooling of buildings in hot arid climates. *Building and Environment*, 43(1):82–89, 2008.

Kolokotsa, D., C. Diakaki, S. Papantoniou, and A. Vlissidis. Numerical experimental analysis of cool roofs application on a laboratory building in Iraklion, Crete, Greece. 2012.

Kruger, K. E., E. Suzuki, and A. Matoski. Evaluation of a Trombe wall system in a subtropical location. *Energy and Buildings*, 66:364–372, 2013.

Li, D. H. W. and J. C. Lam. Evaluation of lighting performance in office buildings with daylighting controls. *Energy and Buildings*, 33:793–803, 2001.

Li, H., Y. Yuebin, N. Fuxin, S. Michel, and C. Bing. Performance of a coupled cooling system with earth-to-air heat exchanger and solar chimney. *Renewable Energy*, 62:468–477, 2014.

Liu, K. and B. Baskaran. Thermal performance of green roofs through field evaluation. National Research Council. Proceedings for the First North American Green Roof Infrastructure Conference, Awards and Trade Show, Chicago, IL, 1–10, 2003.

Maerefat, M. and A. P. Haghighi. Passive cooling of buildings by using integrated earth to air heat exchanger and solar chimney. *Renewable Energy*, 35(10):2316–2324, 2010.

Mahlia, T. M. I. and A. Iqbal. Cost benefits analysis and emission reductions of optimum thickness and air gaps for selected insulation materials for building walls in Maldives. *Energy*, 35(5):2242–2250, 2010.

Mahlia, T. M. I., B. N. Taufiq, K. P. Ong, and R. Saidur. Exergy analysis for day lighting, electric lighting and space cooling systems for a room space in a tropical climate. *Energy and Buildings*, 43(7):1676–1684, 2011.

Nabil, A. and J. Mardaljevic. Useful daylight illuminances: A replacement for daylight factors. *Energy and Buildings*, 38:905–913, 2006.

Nayak, S., K. Ajit, M. Jaya, and G. N. Tiwari. Drying and Testing of mint (Mentha Piperita) by a Hybrid Photovoltaic-Thermal (PVT) based Greenhouse Dryer. *Drying Technology*, 29(9):1002–1009, 2011.

Ozel, M. Determination of optimum insulation thickness based on cooling transmission load for building walls in a hot climate. *Energy Conversion and Management*, 66:106–114, 2013.

Pacheco, R., J. Ordonez, and G. Martinez. Energy efficient design of building: A review. *Renewable and Sustainable Energy Reviews*, 16:3559–3573, 2012.

Parker, D. S. Theoretical evaluation of the night cool nocturnal radiation cooling concept. FSEC-CR 1502-05. Florida Solar Energy Center: Contract Report, 44, 2005.

Peretti, C., Z. Angelo, C. M. De, and Z. Roberto. The design and environmental evaluation of earth-to-air heat exchangers (EAHE). A literature review. *Renewable and Sustainable Energy Reviews*, 28:107–116, 2013.

Pisello, A. L. and F. Cotana. The thermal effect of an innovative cool roof on residential buildings in Italy: Results from two years of continuous monitoring. *Energy and Buildings*, 69:154–164, 2014.

Reyes, V. A., S. L. Moya, J. M. Morales, and F. Z. Sierra-Espinosa. A study of air flow and heat transfer in building-wind tower passive cooling systems applied to arid and semi-arid regions of mexico. *Energy and Buildings*, 66: 211–221, 2013.

Roche, L., E. Dewey, and P. Littlefair. Occupant reactions to daylight in offices. *Light Res Technol*, 32(3):119–26, 2000.

Saadatian, O., K. Sopian, C. H. Lim, A. Nilofar, and M. Y. Sulaiman. Trombe walls: A review of opportunities and challenges in research and development. *Renewable and Sustainable Energy Reviews*, 16(8):6340–6351, 2012.

Santamouris, M., A. Synnefa, D. Kolokotsa, V. Dimitriou, and A. lakis K. Passive cooling of the built environment — use of innovative reflective materials to fight heat island and decrease cooling needs. *Int. J. Low CarbonTechnol*, 3(2):71–82, 2008.

Sherif, A. H., H. M. Sabry, and M. I. Gadelhak. The impact of changing solar screen rotation angle and its opening aspect ratios on daylight availability in residential desert buildings. *Solar Energy*, 86:3353–3363, 2012.

Singh, U., A. Kumar, A. Srivastava, and G. N. Tiwari. Passive cooling of building using multi air gaps. *Applied Energy*, 9(1):33–48, 1981.

Sodha, M. S., S. C. Kaushik, G. N. Tiwari, I. C. Goyal, M. A. S. Malik, and A. K. Khatry. Optimum distribution of insulation inside and outside the roof. *Building and Environment*, 14(1):47–52, 1979a.

Sodha, M. S., A. Srivastava, and G. N. Tiwari. Thermal performance of double hollow wall/roof. *Energy Research*, 3:349–356, 1979b.

Sodha, M. S., A. Kumar, S. Usha, and G. N. Tiwari. Periodic theory of an open roof pond. *Applied Energy*, 7:305–319, 1980a.

Sodha, M. S., A. Srivastava, A. Kumar, and G. N. Tiwari. Heating/cooling of building by flow of water over the roof. *Applied Energy*, 7(13):229–242, 1980b.

Spala, A., H. S. Bagiorgasa, M. N. Assimakopoulosb, J. Kalavrouziotisa, D. Matthopoulosa, and M. G. On the green roof system. selection, state of the art and energy potential investigation of a system installed in an office building in Athens, Greece. *Renewable Energy*, 33:173–177, 2008.

Stazi, F., A. Mastrucci, and C. D. Perna. Trombe wall management in summer conditions: An experimental study. *Solar Energy*, 86:2839–2851, 2012.

Stevanović, S. Optimization of passive solar design strategies: A review. *Renewable and Sustainable Energy Reviews*, 25:177–196, 2013.

Sudan, M. and G. N. Tiwari. Energy matrices of the building by incorporatin daylight concept for composite climate — An experimental study. *Journal of Renewable and Sustainable Energy*, 6:053122, 2014.

Synnefa, A., M. Santamouris, and H. Akbari. Estimating the effect of using cool coatings on energy loads and thermal comfort in residential buildings in various climatic conditions. *Energy and Buildings*, 39:1167–1174, 2007.

Tang, R. and Y. Etzion. On thermal performance of an improved roof pond for cooling buildings. *Building and Environment*, 39(2):201–209, 2004.

Tiwari, G. N. *Solar Energy: Fundamental, Design, Modelling and Applications*. Narosa Publishing House, New Delhi, 2002.

Tiwari, G. N. *Solar Energy Technology Advances*. Nova Science Publishers, Inc, New York, 2006.

Tiwari, G. N. and R. K. Mishra. Advanced renewable energy sources. Royal Society of Chemistry (RSC), (UK), 2011.

Tiwari, G. N., U. Singh, and M. S. Sodha. Roof cooling by water evaporation. *Reviews of Renewable Energy Sources*, 2:148–229, 1983.

Tiwari, G. N., R. K. Mishra, and S. C. Solanki. Photovoltaic modules and their applications: A review on thermal modelling. *Applied Energy*, 88(7): 2287–2304, 2011.

Tiwari, G. N., S. Vikram, S. Joshi Poonam, P. Deo Arjun, and G. Agam. Design of an earth air heat exchanger (EAHE) for climatic condition of Chennai, India. *The Open Environmental Sciences*, 8:18–28, 2014.

Trombe, F. Us patent. 3:832–992, 1972.

Tummu, P., C. Surapong, H. V. Duc, C. Pipat, and R. Pattana. Thermal performance of insulated walls enclosing residential spaces in thailand. *Energy and Buildings*, 61:323–332, 2013.

Tyagi, V. V. and D. Buddhi. PCM thermal storage in buildings: A state of art. *Renewable and Sustainable Energy Reviews*, 11:1146–1166, 2007.

Wang, Y., H. Zhiye, and H. Li. Cost-effectiveness assessment of insulated exterior walls of residential buildings in cold climate. *International Journal of Project Management*, 25(2):143–149, 2007.

Wikipedia. https://en.wikipedia.org/wiki/Embodied_energy.

Wikipedia. https://en.wikipedia.org/wiki/Trombewall.

Wong, I., H. L. Choi, and H. Yang. Simulation and experimental studies on natural lighting in enclosed lift lobbies of high rise residential buildings by remote source solar lighting. *Applied Energy*, 92:705–713, 2012.

Wright, J. and K. Letherman. Illuminance in atria: review of prediction methods. *Lighting Research & Technology*, 30(1):1–10, 1998.

Zhang, Z. L. and B. J. Wachenfeldt. Numerical study on the heat storing capacity of concrete walls with air cavities. *Energy and Buildings*, 41(7): 769–773, 2009.

Zheng, H., C. Feng, Y. Su, R. Wang, and X. Xue. Performance analysis and experimental investigation of a novel trough daylight concentration and axial transmission system. *Solar Energy*, 97:200–207, 2013.

Zinzi, M. and S. Agnoli. Cool and green roofs. an energy and comfort comparison between passive cooling and mitigation urban heat island techniques for residential buildings in the mediterranean region. *Energy and Buildings*, 55:66–76, 2012.

Lightning Source UK Ltd.
Milton Keynes UK
UKHW02f2054110118
315956UK00003B/131/P

9 781680 831962